Biomass and Cell Culturing Techniques

Biomass and Cell Culturing Techniques

Editor
Ramniwas Sharma

2006
Biotech Books
Delhi - 110 035

2006, Publisher
© Reserved

All rights reserved. Including the right to translate or to reproduce this book or parts thereof except for brief quotations in critical reviews.

ISBN 81-7622-161-9

Published by : **BIOTECH BOOKS**
1123/74, Tri Nagar,
DELHI – 110 035
Phone: 27382765
e-mail: biotechbooks@yahoo.co.in

Showroom : 4762-63/23, Ansari Road, Darya Ganj,
NEW DELHI - 110 002
Phone: 23245578, 23244987

Printed at : Tarun Offset Printers,

PREFACE

Cell culturing has been practiced for thousands of years. Domestication, classical plant breeding and genetic engineering are all processes that alter the genome of a plant to enhance its qualities as a crop. Cell culturing is practiced worldwide by government institutions and commercial enterprises. International development agencies believe that breeding new crops is important for ensuring food security and developing practices of sustainable agriculture through the development of crops suitable for their environment.

Biomass is organic non-fossil material, collectively. In other words, biomass is all plant and animal matter on the Earth's surface. In many ways biomass can be considered as a form of stored solar energy. The energy of the sun is 'captured' through the process of photosynthesis in growing plants. Biomass is sometimes burned as fuel for cooking and to produce electricity and heat. This is called Biofuel. Biomass used as fuel often consists of underutilized types, like chaff and animal waste. This is often considered a type of alternative energy, although it is a polluting one.

Editor

Contents

Chapter 1

Cell Organisation

Microscopic Techniques

The application of video technology and computer-aided analysis of images has allowed studies of cell structure and function that were not possible before. Movie and video-tape recording of microscopic data is being superseded by digital image acquisition. This technique is convenient, fast and the duplication of the images does not lose information. Application of digital image technology has specific computer requirements, no longer difficult to obtain. Furthermore, a video camera must be interfaced to a computer through an analog to digital board.

However, digital cameras are also available requiring a camera interface board. Digital to video conversion can be obtained in a variety of ways such as a scan converter. The practical requirements for this technique has been recently reviewed. Refinements of more conventional approaches, such as cytochemistry and autoradiography have also contributed valuable information. Cytochemistry allows the detection and quantitative estimation of cell components or enzyme reactions, whereas autoradiography detects photographically the presence of radioactive isotopes incorporated into cell components. Similarly, immunological techniques have allowed recognition of the distribution of specific proteins inside cells. In situ hybridization, which allows recognition of specific nucleotide sequences, identifies the location of nucleic acids.

Light Microscopy

Fig. 1 illustrates the principles of two of the light microscopic methods currently used in cell biology: conventional light microscopy (Fig. 1a) and phase and interference microscopy (Fig. 1b).

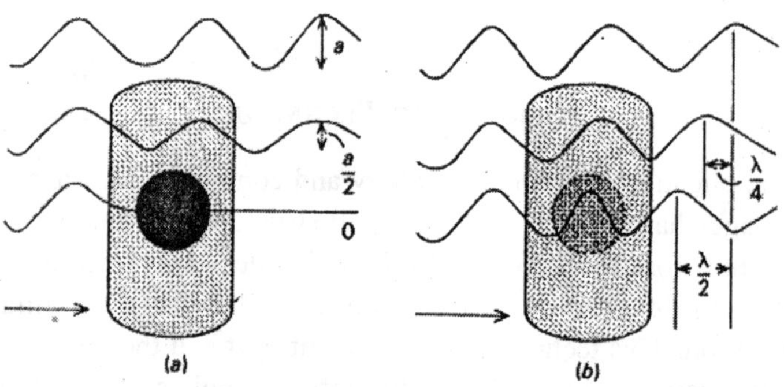

Figure 1: Comparison of absorption and phase interference methods in light microscopy.

With visible light and an ordinary light microscope, the various cell components offer little contrast unless they have been stained or absorb light. Generally, before being examined with the light microscope, a specimen must be fixed, a procedure that immobilizes the macromolecular components either by denaturing or cross-linking them. Then embedding in either plastic or paraffin is needed, generally preceded by dehydration. The material is then sectioned and stained. The absorption or fluorescence of the dyes can be measured quantitatively with a variety of techniques. Fig. 1a illustrates the absorption of light through a stained cell in conventional microscopy.

In light microscopy, immunological techniques have been used by attaching an antibody to a label, a dye with characteristic light absorption or fluorescence. Similarly, sites of enzyme activity can be identified by using substrates, sometimes analogs of the natural substrates, which the enzymes convert into insoluble compounds with characteristic absorption or compounds that can be readily precipitated. The interference methods illustrated by the special case of phase-contrast microscopy in Fig. 1b do not require any of the preparatory procedures, so they are best suited for the study of living cells whose dynamics can be recorded on videotape. Many past contributions were made with cinematography. These methods take advantage of light interference phenomena.

When two coherent beams of light are recombined, they can interfere with each other constructively or destructively to produce a contrast image. The beam passing through an object will be retarded in proportion to the thickness and density of the object. In phase-contrast microscopy the beam diffracted by the object is recombined with a reference beam that has passed through the medium or the background material. The reference beam has been advanced or retarded a quarter-wavelength to maximize the interference. The difference in phase between the two beams will give positive or negative interference, the former showing the object as darker and the latter as lighter than the background. In interference microscopy, the phase of the reference beam can be varied in relation to the specimen beam, allowing measurement of the retardation of the specimen beam. The method can be used as a quantitative tool to estimate the dry weight of objects.

Nomarski differential interference contrast microscopy (DIC) is based on the interference between two closely separated points in the object. The beam passing through the specimen is split by a birefringent plate. The image is a gradient of the phase difference between these two adjacent points. Mathematically, the contrast is the derivative of the path differences with respect to distance. The image has a directional contrast resembling a shadow-cast relief map of cellular details. This technique provides a resolution closer to the theoretical limit than any other light microscopic technique. Furthermore, objects above and below the plane of focus are excluded from the image, which provides essentially an optical section.

Nomarski optics are ideally suited for observing objects with well-defined boundaries, such as fibers or condensed chromosomes. Less powerful techniques, such as modulation contrast microscopy and other asymmetric illumination methods, provide images that resemble DIC images. Hoffman microscopy has been used recently because of its economy and its advantages in common with DIC, such as viewing of optical sections, increased visibility, and contrast of unstained specimens. As in DIC, the images are shadowed and show no halos. However, the principles involved are very different.

Fig. 2 allows a comparison of images obtained with phase-contrast and Nomarski differential interference microscopy. With Nomarski microscopy (Fig. 2a) the details are sharp and each cell component is clearly defined. With phase contrast (Fig. 2b with positive contrast) the details appear more diffuse and, where slightly out of focus, are surrounded by halos. Generally, under optimal conditions, light microscopy can distinguish as separate images two points that are no less than 0.2 μm apart. This is considered the practical limit of resolution for these techniques. With shorter wavelengths, much closer points can be resolved; this principle has been exploited in electron microscopy, which uses electrons instead of visible light. The resolution with the electron microscope under optimal conditions approaches atomic dimensions.

Fig 2 showing, Endosperm cells of Hemanthus katherinae shown with (a) Nomarski differential interference. Type some more Endosperm cells of Hemanthus katherinae shown with (b) positive phase contrast. The bar represents 20 mm. Although light microscopy is limited in resolution, objects below the limits of resolution can still be studied using Nomarski optics and high-performance video cameras. The size of the image, however, does not correspond to that of the object. The video methods enhance the contrast of the objects by subtracting the background and amplifying the remaining signal.

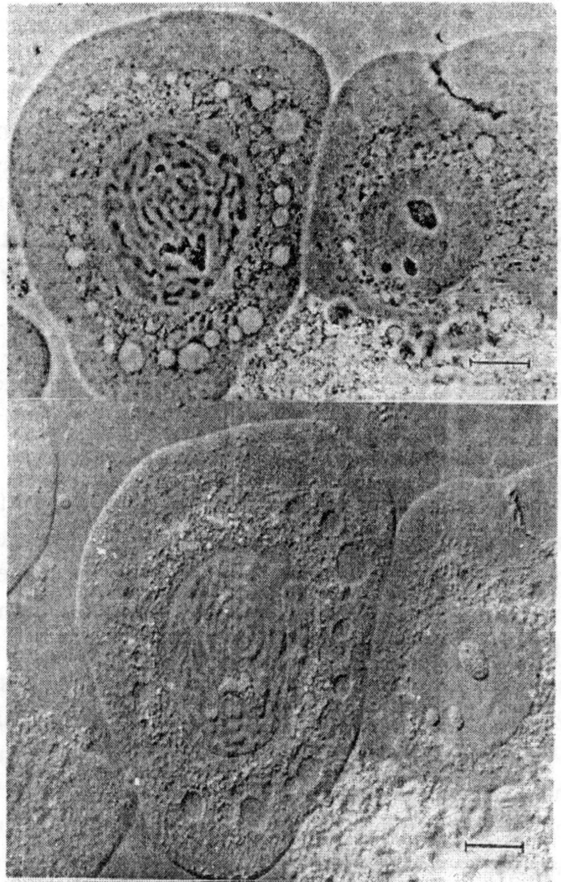

Figure 2: Endosperm cells of Hemanthus katherinae

This approach allows viewing of single microtubules that are only 25 nm in diameter. Contrast improvements and light intensity measurements have also been provided by digital image processors that filter and average signals. Confocal microscopes for imaging either fluorescence emission or reflected light have been developed and are now commercially available. Like DIC, they allow the study of optical sections.

The instruments contain a computer-controlled system that scans the specimen rapidly in three dimensions with a single small laser illuminated spot. Light from the specimen passes through an aperture and is focused on the plate of a video camera. Unlike conventional microscopes, they view a region coincident with the illuminated spot. Only the region lying within a narrow depth of focus produces an image that corresponds to a precise location in space. Ideally, out-of-focus areas are not a problem since they are excluded by the pinhole aperture. The image can be digitized and stored on computer disks for analysis with conventional programs. Confocal microscopes are ideal for studying the three-dimensional organization of cells with fluorescent dyes. Confocal microscopy is a powerful technique capable of providing a wealth of 3-D information on a living specimen.

However, the entire specimen is exposed to the light producing bleaching and photodynamic damage mediated by the fluorophore or other absorbing molecules. Because only a fraction of the emitted light is used, the amount of light needed is substantial with subsequent increase in damage. In addition, with thick specimen the information is restricted to structures about 20 μm from the surface because of distortions caused by scatter. Most of these problems can be avoided by the use of multiphoton optical absorption to mediate excitation as done with two photon-fluorescence microscopy. Molecules can absorb simultaneously two or more photons delivered by intense laser pulses. The high intensity laser used has a short duration and at 100 MHz it boosts the frequency of two photon absorption. The exciting light, in the infrared range, is not absorbed from single photon eliminating background fluorescence and background scattering.

The fluorescence is generated virtually only in the vicinity of the focal point. In scanning, fluorescence excitation is limited to a focal "slice". In addition, all fluorescence photons are used. It has been found possible to make observations on very thick specimen. Evanescent field fluorescence microscopy and the availability of novel fluorescent substances has opened new avenues in the study of events occurring in or close to plasma membranes. When a beam of light passes from a medium of high refractive index to that of a lower refractive index the light undergoes total internal reflection if the angle of incidence is sufficiently high. This reflection produces a thin layer of light, the evanescent field (EF). Consequently molecules in this thin layer are illuminated, whereas molecules outside this zone remain in the dark.

In addition to exciting fluorescent molecules in the thin illuminated layer, the technique provides information on the position of the fluorescent object which brightens as it approaches the interface and dims when it leaves it. An entirely new concept of microscopy developed by M. Isaacson and A. Lewis promises to extend the resolving power of the light microscope to as little as 50 nm. This microscope scans the illuminated specimen in 15-nm steps with a tiny detector consisting of a tube with an aperture as small as 50 nm in inner diameter. 3D reconstruction has allowed the mapping of gene expression patterns in development using light methods. A variety of techniques have been developed for 3D reconstruction of large objects such as embryos.

The problems are somewhat different from conventional microscopy because of the large size of the specimen. Furthermore, these techniques do not require the high resolution needed for the study of cells. Confocal microscopy allows making observations in specimen as thick as 1 mm. Optical coherence tomography (OCT) can use specimen as thick as 2 to 3 mm by measuring backscattered infrared light and can be used with living specimen. The technique is similar to radar except that in this case backscattering or reflections of light rather than sound are detected at various specimen depths. Optical projection tomography (OPT) has been developed to produce 3D images of stained or fluorescent labelled prepared biological specimens with a thickness of up to 15 mm. With this technique, the specimen is rotated 360 degrees at angular steps of 0.9 degrees leading to a reconstruction of the image.

Electron Microscopy

The electron microscope (EM) depends on the fact that electron beams can be bent and focused by electric or magnetic fields, allowing them to form magnified images in much the same way as light in light microscopy. Electron microscopy has brought the study of cell structure to the so-called submicroscopic or ultrastructural level, where even macromolecules can now be observed. The electron beam originates from a filament and, after passing through an evacuated microscope column, is focused on a fluorescent screen or a photographic plate.

Transmission Electron Microscopy: With the transmission electron microscope (TEM), the electron beam passes through the specimen. The denser areas in the specimen scatter more electrons and the corresponding regions in the image appear darker. Electron stains of high density are usually used to enhance the contrast. The specimen preparation generally used for TEM resembles in principle the fixation, embedding, sectioning, and staining used for the light microscope. Because of the increased magnification and spacial resolution offered by EM, specimen preparation is especially critical. Ideally, in order to obtain an accurate image of their structure, the components of cells and tissues must be maintained in a configuration corresponding to their original ultrastructure.

Many fixatives such as OsO_4, also serve as stains. Others, such as glutaraldehyde, do not add to the contrast and additional treatment with electron stains is needed. Generally, the sections have to be much thinner than those used for light microscopy, which must be embedded in plastic. In part, the thinner sections must be used because of the limited penetrating power of conventional TEM; in part, they are needed to avoid the superposition of features in the image above or below the structures of interest, because through focus optical sectioning or confocal microscopy is not possible in EM. High-voltage or intermediate-voltage electron microscopes are now in use. Because of the increased penetration of the electrons provided by the higher voltages, thicker specimen, sometimes even intact cells, can be used.

Together with tilting of the EM stage, these techniques provide stereoscopic information and a wealth of detail. However, because of the overlap of the information in these images, they are difficult to interpret. Developments in computer analysis of such images are providing greater information about the three-dimensional organization of cells. Isolated cell components, macromolecules, and macromolecular assemblies such as ribosomes and viruses have been studied with the TEM using negative staining. With this technique the particles are suspended, generally in a solution of an electron-dense compound. The preparation is sprayed or spread in some manner on a grid and then dried. Structures that are not permeated by the salt appear light, whereas those containing the salt, for example, the previously hydrated areas-appear dark.

In figure 3, (a) Electron micrograph of a negatively stained mitochondrial outer membrane from the fungus Neurospora crassa (scale bar, 100 nm). (b) The two-dimensional repeat motif of one of the membrane arrays. (c) Optical diffraction of micrograph taken with coherent illumination from a laser. (d) Three-dimensional reconstruction of the array. (e) The projected density of the same membrane array embedded in vitreous ice without stain.

The scanning transmission electron microscope (STEM) has been used to study the structure of the freeze-dried ciliary and flagellar protein dynein without the use of stain or fixative. This technique, which uses a very low electron dosage, also allows the determination of mass from the electron scattering intensities over a particle.

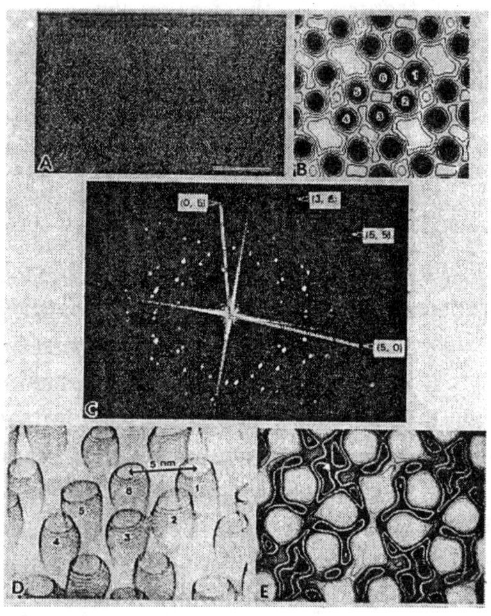

Figure 3: Electron crystallography of a membrane channel.

The newest advance in TEM is the imaging of cells and macromolecules that are embedded in amorphous ice at very low temperatures. These specimens are unfixed and unstained, so the contrast arises directly from differences in density of the cell components. Instrumental or computer techniques are usually needed to enhance the low inherent contrast of these frozen-hydrated images (Fig. 3e).

The use of electron cryomicroscopy together with the development of computational techniques is beginning to approach the level of resoluti sed below in relation to computer-aided reconstruction.

Scanning Electron Microscopy (SEM)

In the scanning electron microscope (SEM) a beam of electrons is focused to form a small diameter probe that is scanned across the specimen in a process similar to that used to produce television images. This probe interacts with atoms in the specimen causing them to emit a variety of signals. The microscope collects and displays electrons scattered from the surface of the object. Generally, the specimen has to be dried and coated with metal. The metal increases the electrons scattering and also acts as a conductor to avoid accumulation of charges. The SEM produces images with a stereoscopic appearance because curved portions of the object reflect more electrons than flat areas. The SEM is very useful in observing the surfaces of three-dimensional objects.

However, it is typically restricted in resolution to above 5 to 10 nm. The principles involved in forming an image are shown in Fig. 4. Secondary electrons are produced from sample-beam interaction but their escape from the specimen and subsequent detection depends on the surface geometry of the specimen as shown in Fig. 4. More escape from projections than from flat surfaces. The differences in the resulting voltage levels in the detector produce the image.

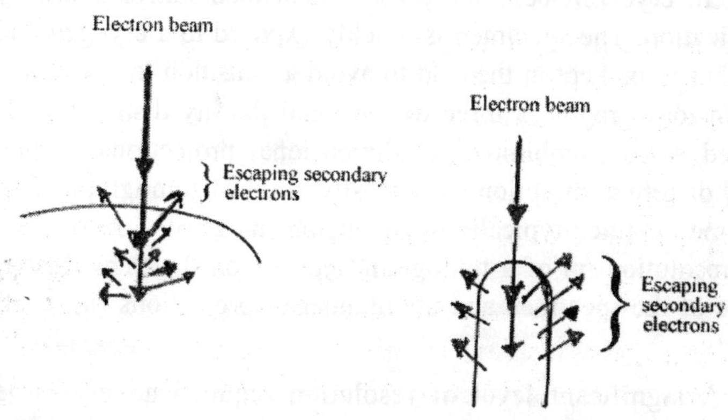

Figure 4: Secondary electron escape, edge effects and image production.

More recently, high-resolution SEM (HRSEM) has been used as an effective tool to study subcellular detail. These instruments approach the resolution of TEM by rastering the specimen with a very small, but extremely intense spot, about 1,000 times brighter than conventional sources. With this technique the specimen is placed within the magnetic field of the final lens. The beam interacts with the surface of the observed object and the signal of secondary electrons can be collected efficiently with a minimum of aberration.

The voltages used are relatively low. As the beam penetrates only the top 10 nm of the specimen, the image is restricted to a shallow portion of the surface. The material is either fixed or rapidly frozen, the water is generally removed or substituted. Although uncoated specimens can be used, generally they are coated with a metal coat, optimally chromium or tantalum at a thickness of 1 to 12 nm.

Cryo-electron Tomography

Electron cryo-electron tomography holds a promise of reconstructing the actual structure of specimen in their native state. In cryo-EM cells are preserved in their native states by vitrification. The specimen is quickly exposed to a cryogen and then has to be kept in the cold to avoid a transition to crystalline ice. In tomography, a three dimensional density distribution is arrived at by combining two-dimensional projections viewed from different directions Generally, in EM-tomography the specimen is tilted typically in the angular range of -70° to +70°. The resolution (r) of a tomogram depends on the thickness (t) of the of the specimen and the number of projections (n): r~pt/n.

A significant level of resolution requires a very large number of projections recently made possible by the automated collection of tomographic tilts series at low enough electron dosages. The reconstruction of of a ribosome is shown in the cover graphic at the beginning of this textbook. Tomography is more direct for periodically repeating units, such as those in two-dimensional crystals formed in membranes, which allow the application of crystallographic methods. The images of untilted and tilted specimens are first digitized. The three-dimensional reconstruction of the two-dimensional crystal is done with Fourier transforms (FTs). The three-dimensional FT of a two-dimensional crystal is an array of lines normal to the plane of the crystal (Fig. 5). Variations in the specimen density in the z direction in real space result in modulation of intensity along each z line.

The two-dimensional FT of a projection image of the crystal represents a central section, that is, a slice through the origin of the three-dimensional FT with the angle of the slice equal to the tilt angle of the specimen in the EM. By combining many projection transforms, the three-dimensional Fourier volume is filled in, except for a missing cone defined by the maximum tilt angle. The projection transforms are brought to a common phase origin and the inverse transformation is performed, yielding the three-dimensional volume in real space. Fig. 3a shows an electron micrograph of a two-dimensional crystal produced by concentrating the proteins in an outer membrane preparation by partially removing the lipid with phospholipase. The optical diffraction pattern from the micrograph is shown in Fig. 3c. The crystals are formed by the so-called mitochondrial porin or voltage-dependent anionic channel (VDAC). The pores are black when filled with stain (Fig. 3b) and white when filled with water (Fig. 3e). Fig. 3d is a three-dimensional reconstruction of the negatively stained crystals showing the channel structure.

Unordered molecules on an EM grid, which are viewed generally with negative staining or after freezing pose a different challenge, since the analysis is necessarily much more complex. The EM images are recorded with and without tilting of the specimen to provide three-dimensional information, as done for the two-dimensional crystals. The computer three-dimensional reconstruction of the nonperiodic structures is done in real space by tomography. Again, multiple projection images are used; these may be a "conical series," as shown in Fig. 6, in which there is a constant large tilt angle and the complete range of azimuthal angles is covered. After careful alignment of the images and prefiltering in Fourier space to enhance high-resolution details, the reconstruction requires back-projecting the density at each point in each image into appropriate strips in the three-dimensional volume.

A flow diagram of the reconstruction procedure is shown in Fig. 8. Figure 9a and b show 50s ribosomal subunits tilted by 50 and untilted, respectively. A reconstruction of the ribosomes and various ribosomal subunits is shown in Fig. 10.

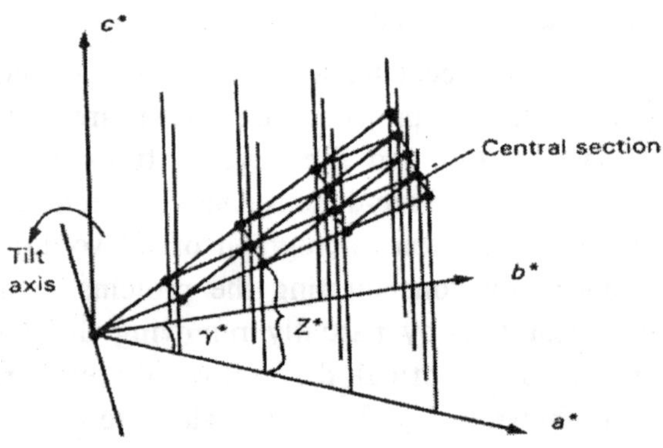

Figure 5: Schematic diagram showing the transform of a two-dimensional crystal.

The transform takes the form of a number of lattice lines extending perpendicular to the plane of the crystal. Each micrograph contains an image that, when Fourier-transformed, gives the values of amplitude and phase at points along the lattice lines where the central section intersects them.

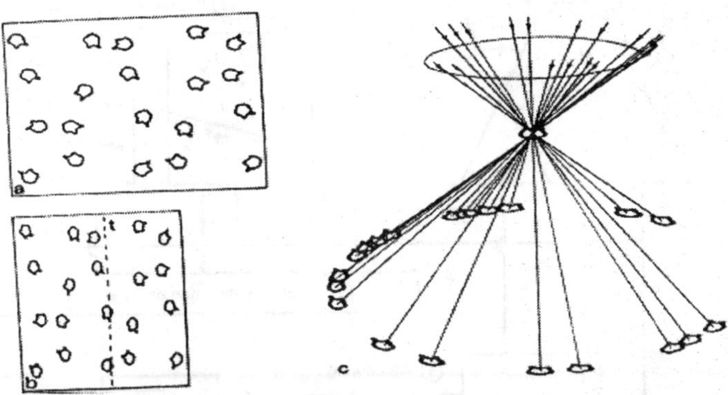

Figure 6: Basic principle of the reconstruction scheme.

In figure 6, (a) View of a specimen with randomly oriented 50s particles lying flat in the plane of the specimen. (b) Projection of the specimen in (a), tilted by 50. The images that can be extracted from the tilted image (b) form the conical tilt series shown in (c), equivalent to a tilt series of a single particle with random projection directions, all lying on the surface of a cone.

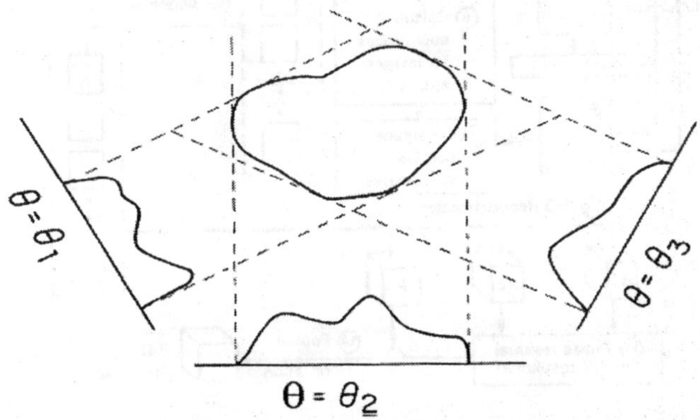

Figure 7: Illustration of the back-projection method of three-dimensional reconstruction.

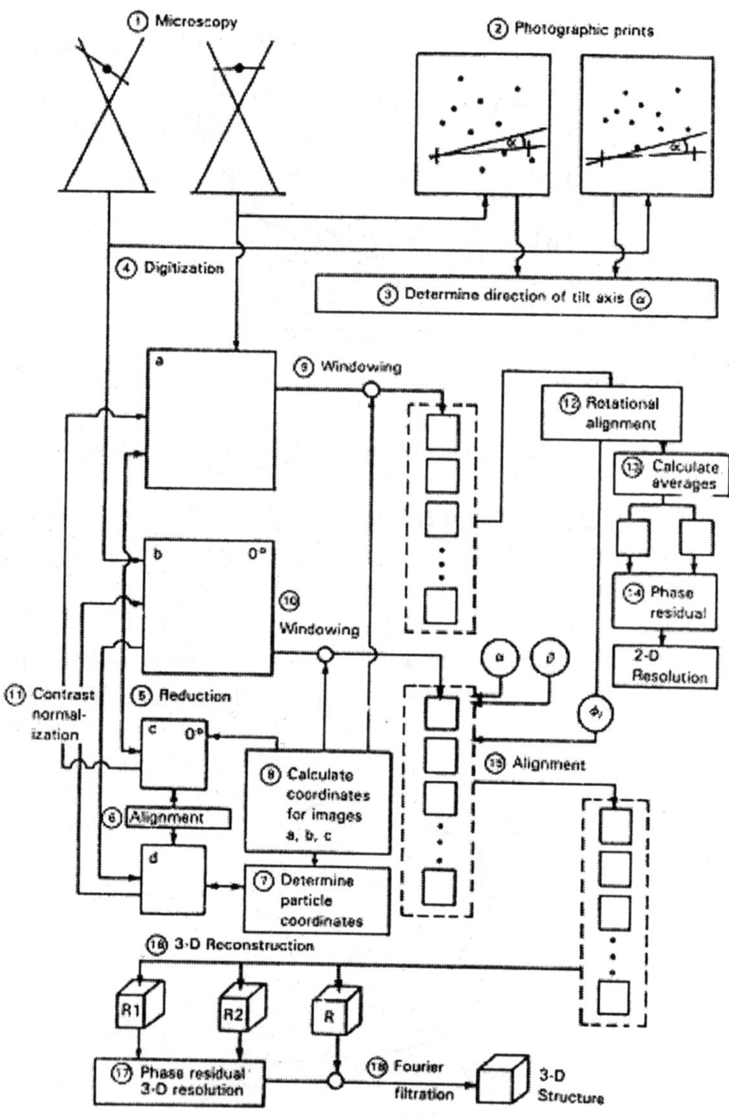

Figure 8: Flow diagram outlining the complete reconstruction procedure.

In any plane perpendicular to the tilt axis, the optical density profiles measured along corresponding lines of the projections are smeared out into the direction of projection and are added up to form a slice of the three-dimensional object to be reconstructed.

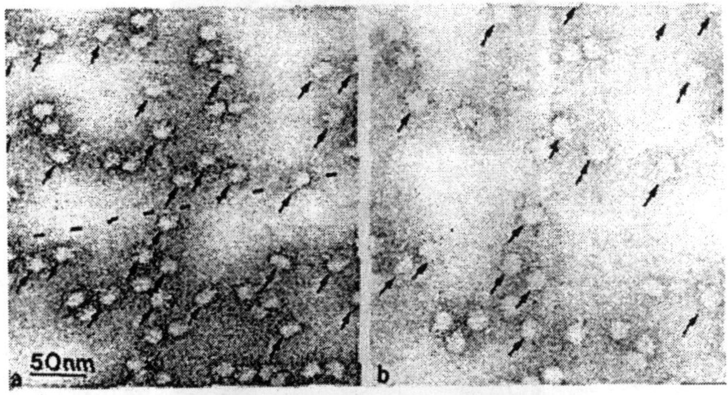

Figure 9: Portion of the particle fields.

In figure 9, (a) Projection of the specimen tilted by 50; the particles initially selected are marked at the lower left corner. (b) The same field without tilt; the particles finally used are marked. Only the area common to both micrographs and lying in the underfocus range in the tilted view (as determined by optical diffraction) was used for evaluation to avoid errors due to the sign change of the transfer function at low spatial frequencies. Average densities calculated in areas showing only carbon foil were used to normalize the contrast of the images.

Figure 10 shows the Gallery of ribosomes and ribosomal subunits reconstructed with the new technique, presented as stereo pairs. (a) 50S ribosomal subunit of E. coli, oriented with its interface surface toward the viewer. (b) 50S ribosomal subunit of E. coli depleted of L7/L12 proteins. Orientation as in (a). (c) 705 monosome of E. coli oriented so that the 505 domain matches position with (a) and (b). (d) 405 ribosomal subunit from rabbit reticulocytes in a lateral-view orientation.

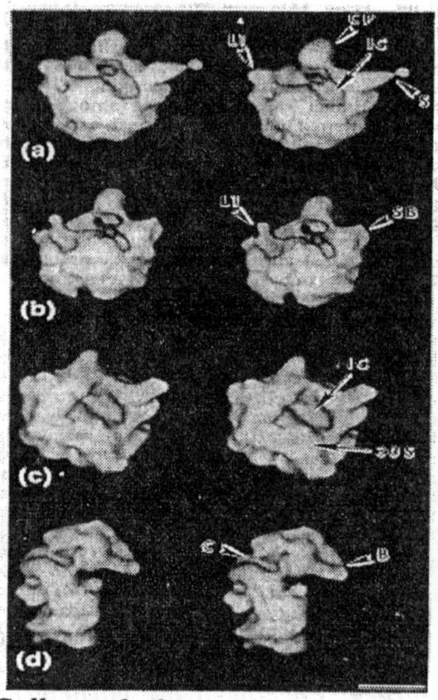

Figure 10: Gallery of ribosomes and ribosomal subunits

Scanning Probe Microscopy

The scanning tunneling microscope (STM) has been applied to the study of biological surfaces to a very limited extent, since it is a relatively new application. However, its potential is considerable since it allows observations with atomic resolution. The method requires very thin molecular layers attached to a conductive substrate or alternatively coated with a conducting surface. A metal tip is brought close to the surface and scans at moderate voltages, in the range of 2 mV to 2 V. Electrons tunnel between the tip and the conducting surface. A feedback system maintains the current constant by changing the height of the tip. The image is a map of the tip height in relation to the lateral position of the probe. The resolution is generally very good because invariably some small protuberance projects out of the tip giving atomic resolution.

The atomic force microscope (AFM) can be used for imaging as well as micromanipulation. The AFM scans the surface with a sharp tip attached to a soft cantilever. Deflections of the tip, which correspond to the surface topography are recorded optically from the cantilever displacement. Various recent technologies have been used for the fabrication of the necessary devices. To date, the resolution of the system using biological specimen is in the range of 1 to 50 nm.

Cell Components Isolation

The isolation of cell components has also led to enormous progress in our understanding of the molecular organization of cells. Basically, this approach depends on rupturing the cell in a medium that is compatible with preservation of the integrity of the organelle under study, followed by physical separation of the various cell components. Generally, the medium is a solution that is isosmotic or hyperosmotic in relation to the cell interior; chilled sucrose solutions are often used. Cells and tissues are usually disrupted mechanically. Bacteria and plants present unique problems because of their tough protective capsules or walls. The capsules or walls can be broken down by enzymatic digestion, or can be prevented from forming by use of special techniques such as using a mutant with a defective capsule. Without such coverings, homogenization can be relatively gentle.

For example, with the homogenization device shown in Fig. 11, up-and-down motion of the plunger forces the cells to squeeze through the small space between plunger and vessel wall. The resulting shear breaks down the plasma membrane and some of the cytoskeleton. The homogenizer can be maintained in a ice bath to dissipate the heat generated and to minimize the effect of hydrolytic enzymes. The most widely used methods for large-scale separations involve centrifugation. The homogenate is placed in a plastic or heavy-walled test tube and then centrifuged. The rotation generates a centrifugal force. which increases with the speed of rotation and the radius of the rotor. The rate of sedimentation of a spherical particle is given by Stokes' law, shown in Eq. (1):

$$\frac{dx}{dt} = \frac{KG(d_p - d_m) r^2}{\eta} \quad (1)$$

In this equation, x is the displacement in cm, *t* is time in seconds, *G* is the force exerted on the particle in gravitational units, d_p, is the density of the particle (g/cm³) and d_m that of the medium, *r* is the radius of the particle in cm, and is the viscosity of the medium in poises, *K* corresponds to 2/9. The centrifugal force is a function of the rotations per minute (rpm) of the rotor as represented in Eq. (2), where L is the radius of the rotor in cm.

$$G = 11.17L \frac{rpm}{1000}^2 \quad (2)$$

Equation (1) shows that particles differing in *r* and d_p, could be isolated by sedimenting them sequentially increasing the speed with each centrifugation, as done in differential centrifugation. In contrast, when d_p, is lower than the density of the medium ($d_p < d_m$), *dx/dt* would be negative and the particle would he displaced upward.

In differential centrifugation, after each centrifugation, samples are separated into distinct fractions: a pellet of packed sedimented particles and a supernatant containing unsedimethed material. The larger or denser particles sediment in earlier centrifugations. This process is illustrated in Fig. 11. Differential centrifugation cannot separate cell components into pure fractions in a single series of centrifugations, because particles will sediment according to their position in the tube rather than just size or density. However, repeated resuspensions and resedimentations can produce a satisfactory level of purity.

The effect of position in the tube on isolation can be avoided by layering the homogenate on top of a layer of a denser solution. This process approximately equalizes the distance of travel for all particles. Layers or gradients can be produced by varying the concentration of sucrose or of a polymer in the layers, with the denser layers in the bottom.

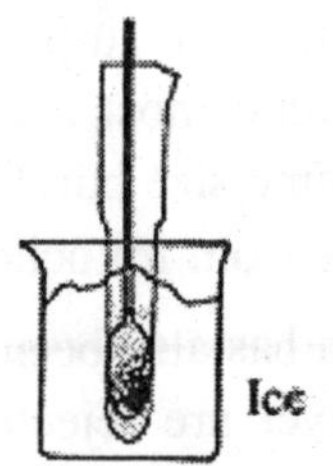

Figure 11: Diagrammatic representation of differential centrifugation.

Density gradient centrifugation can provice high resolution. As a particle approaches a layer of identical density ($d_p = d_m$) the rate of sedimentation decreases [Eq. (1)] and eventually, at equilibrium, sedimentation stops. With equilibrium centrifugation the separation will depend exclusively on the density, not the size of the particle. However, in many cases the centrifugation need not be carried out until equilibrium. Not surprisingly, density gradient centrifugation has been found most useful for separating organelles of approximately the same size but different densities for example, separating lysosomes from mitochondria.

Centrifugation with a two-phase system has also been found useful. Particles lighter than the bottom layer are rejected and remain at the interface. Rapidly spinning rotors generate considerable heat, therefore centrifugation techniques require refrigeration. In addition, to reach high speeds the friction has to be decreased by use of a partial vacuum around the rotor. Subcellular particles can be separated by taking advantage of their surface properties. Many aqueous solutions containing two polymers, such as methylcellulose and dextran, separate out into phases. If the solutions have been mixed with subcellular components, viruses, or macromolecules, these substances will tend to favor one of the two phases or the interface between the two.

Organelles and other particles can also be sorted out by flow cytometry because a variety of fluorescent dyes stain organelles specifically. With this technique, the flow of medium containing the particles is delivered by a nozzle and broken into individual droplets by a vibrator, each droplet containing a single particle. Depending on the fluorescence, the droplets are given either a positive or negative electrical charge. They are then separated by a strong electric field and collected according to charge. This approach has considerable promise, particularly if used after more conventional techniques, combining bulk isolation with higher precision.

The isolation techniques have made available large quantities of material for study by standard biochemical techniques. However, in some cases, lack of purity and possible disruption of native organelles still pose a problem. The microscopic and biochemical techniques discussed have individual advantages and disadvantages and have been used together to produce a complex picture of cellular organization.

Cells Manipulation

Micromanipulation techniques have a long history. Parts of cells have been removed or moved, dyes or exogenous macromolecules have been injected into cells. Laser beams have been used in microsurgery. Several new techniques have introduced new prospectives to these approaches. There are alternative to to the introduction of components by micronjection into cells. Streptolysin O is pore-forming toxin that can permeabilize cells to proteins as large as 100 kDa. The permabilization can be reversed in the presence of Ca^{2+}-calmodulin and and intact microtubules. Resealed cells were found to be viable and were capable of proliferation and endocytosis.

An entirely different approach is provided by a class of peptides called penetratins. These peptides enter cells freely and can carry with them whatever compounds have been attached to them covalently, even iron nanoparticles 40 nm in diameter! In practice, they can be linked either by chemical reactions or genetic manipulation. Other polypeptides have been shown to have similar properties. Penetratins possess polybasic protein-transduction domains (PTDs) which bind to negatively charged groups. They are throught to interact with lipids to form inverted vesicles with the hydrophilic groups inside and the hydrophobic groups outside.

However, a more recent analysis suggests that the uptake proceeds by endocytosis and binding to heparan sulfate proteoglycans (HSPGs) at the cell surface has been implicated in the uptake. Regardless of mechanism, the use of penentratins is promising and may provide a way of introducing a number of macromolecule into cells. Biologically, among the proteins that are transferred in this manner are homeoproteins. Homeoproteins are transcription factors involved in developmental steps. The transfer of these proteins could transfer positional information between cells during development. Cell-permeable peptides have also been used in the study of protein interactions. The penatrins were linked to intracellular inhibitors, peptides containing protein interaction domains. By competing with the endogenous proteins containing the same domain they block their biological function.

More recently, this approach has been used to define the pathways of export from the nucleus of particular mRNAs, using cell-permeable peptides constructed to interfere with interactions between known adapter and receptor proteins. The use of single beam laser optical gradient traps, also called optical tweezers, holds considerable promise. Optical tweezers allow experimenters to control of position to an accuracy of about 10 nm. They also allow forces in the order of 10^{-8} dynes. Optical tweezers can trap particles ranging in size between 15 nm to 25 μm and thus manipulate macromolecules in solution as well as organelles and whole cells.

Rays of light are refracted at the surface of a particle. Each passing photon, by virtue of its momentum, gives the particle a minuscule "kick" which can displace it. A single Gaussian laser with a Gaussian transverse intensity profile produces a stable configuration close to the beam focus and acts as a trap. Optical tweezers have allowed, for example, measuring minute forces. The principles involved in optical tweezers are depicted in fig.12. In this figure the solid arrows represent the refracted light rays. The open arrows represent the reactive forces produced by conservation of momentum imparting a momentum equal and opposite to the momentum change of the rays.

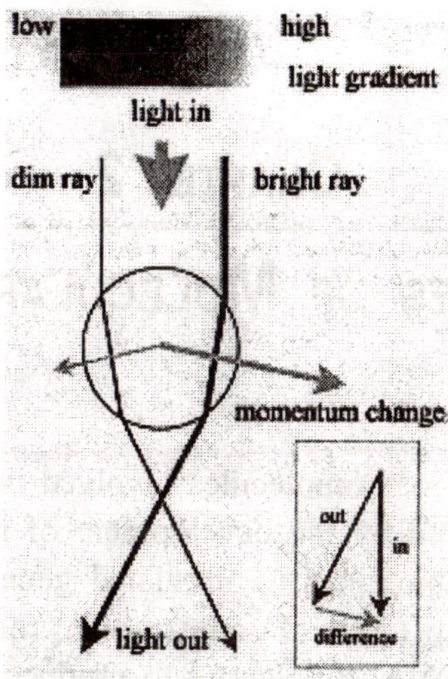

Figure 12: A parallel beam of light with a gradient in intensity impinges on a spherical lens.

Two representative rays are drawn. The rays are bent as shown. As shown by the vector diagram the rays pull the sphere towards the brighter light. In the absence of a gradient the sphere is "trapped".

CHAPTER 2

TECHNIQUES OF MOLECULAR BIOLOGY

The study of the macromolecules involved in cell function has been revolutionized by the development of recombinant DNA technology. Combinations of unrelated genes can be created, cloned and amplified. Furthermore, the production of complementary DNA, its cloning to produce large amounts of material and subsequent sequencing have allowed the study of proteins.

RECOMBINANT DNA TECHNIQUES

Constructing DNA

Novel DNA can be produced by the use of techniques in which a piece of DNA is cut in pieces and then combined with other DNAs. Restriction enzymes or endonuclease that can cut up DNA molecules into smaller pieces are found in a variety of prokaryotes. They recognize specific base sequences in double-helical DNA and cleave them in such a way that the new ends are single stranded and complementary to each other; that is, two complementary strands are cut in slightly different positions, producing staggered ends (Fig. 1a).

The same restriction enzyme can be used to cut vector-DNA, creating staggered ends with precisely the same sequence as the donor DNA. These so called sticky ends allow the restricted fragments to be combined with the vector DNA. Vectors are phages or plasmids independently replicating nonchromosomal DNA. Treatment with DNA ligase will then join the two kinds of DNA, the vector and the donor piece (Fig. 1b). Similarly, a small, artificially synthesized DNA piece can be covalently joined to either vector or donor DNA.

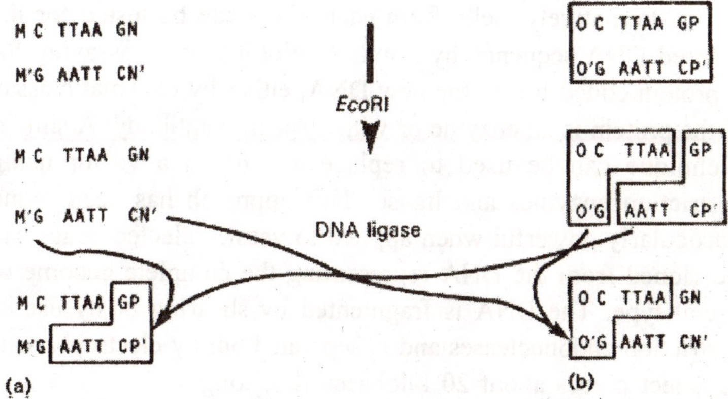

Fiure 1: Production of a novel DNA. Only the staggered ends are shown. (a) Production of staggered ends by endonuclease. (b) Joining of the two distinct DNAs by DNA ligase.

The covalent joining of a DNA fragment to a DNA vector permits cloning of the fragment, that is, the production of many identical copies. DNA can be introduced into cells by a variety of techniques, including infection with a virus containing the donor DNA. The modified DNA replicates in host cells. A commonly used host for cloning is Escherichia coli, the intestinal bacterium that has already provided us with a wealth of genetic and biochemical information. A method has been developed for the delivery of DNA by producing 25nm balls containing compacted DNA and positively charged peptides. These particles appcar to cnter cells readily and will be transferred into the nucleus.

Supposedly, this technique compares favorably with the use of liposomes or virus vectors to transfer the genetic information.

The cells containing the donor DNA can be selected by various procedures. Most simply, the DNA is ligated to a vector containing a marker gene, such as antibiotic resistance. This allows recombinant organisms to be selected by growing cells in the presence of antibiotic where only the cells expressing the marker survive.

Alternatively, cells from each clone can be tested for the desired DNA sequence by Southern blotting or by assaying for a protein coded for by the new DNA, either by enzymatic assay if the protein is an enzyme or with a specific antibody. A similar technique can be used to replace a gene in a vector using restriction enzymes and ligase. This approach has been found particularly powerful when applied to yeast. Selected genes can be cloned from the DNA representing the complete genome of a cell type. The DNA is fragmented by shearing or by use of restriction endonucleases and is separated out by electrophoresis to select pieces about 20 kilobases (kb) long.

When the DNA pieces are linked to vectors, such as lambda phage DNA, and the infective phage is reconstituted in vitro, they can be used to infect E. coli and replicate repeatedly to form a so-called genomic library. A similar technique departing from the mRNA of a cell, followed by the production of cDNA and cloning is referred to as an expression library. Infection and subsequent lysis of E. coli cells on a plate by a dilute suspension of phage produce plaques, each corresponding to the progeny of a single phage.

When a sheet of nitrocellulose is applied to the plate, some of the DNA from lysed cells sticks to the sheet and forms a replica. After denaturation of the DNA with alkali, specific DNA sequences can be identified by hybridization to radioactive DNA or RNA probes. The hybridization sites can be recognized by autoradiography and the corresponding plaques still present in the original plate can be picked out and grown to form millions of DNA clones. Amplification by synthesis of DNA using the polymerase chain reaction (PCR) with a selected DNA serving as a template provides an alternative to cloning.

Single genomic sequences as large as 2 kb have been amplified more than 10 million times with a high-specificity, heat-resistant DNA polymerase extracted from a thermophilic bacterium. A thermostable polymerase is advantageous, since newly synthesized strands have to be separated (generally done by heat) before the next round of replication. A continuous flow PCR system permits very fast amplification. The sample is heated or cooled by flowing continuously and repeatedly over sections of a device where each section is maintained at constant temperature corresponding to melting, primer annealing and primer extension temperatures.

DNA probes can be produced from the corresponding messenger RNA (mRNA). The mRNA can be isolated from cells that produce predominantly a single protein (e.g., reticulocytes. which synthesize hemoglobin). Alternatively, the mRNA can be extracted from polysomes isolated by precipitation with antibodies against the protein in the process of being synthesized. Then complementary DNA, cDNA, can be synthesized using reverse transcriptase and the mRNA as a template. Reverse transcriptase is an enzyme found in retrovirus, where it synthesizes a DNA template complementary to genomic RNA during replication of its genome.

The cDNA can then be cloned as already described or used to produce the corresponding protein. Site-specific mutagenesis has obvious advantages over conventional mutagenesis, since the latter is a random rather than a directed process. Altering single nucleotides in isolated DNA permits synthesis of the modified DNA. In practice, this can be accomplished by first synthesizing a small oligonucleotide piece complementary to a short sector of the modified DNA containing the nucleotide change. The rest of the strand can then be synthesized by DNA polymerase which uses the small oligonucleotide as a primer.

Site-specific mutagenesis has been used to modify proteins in a controlled and systematic manner in the study of mechanisms of catalysis and the relationship between protein structure and function. Homologous recombination between exogenous DNA and a target locus can be used to place foreign DNA at precise sites in the chromosome. Replacing a DNA sector pin-points a mutation precisely (producing a mutation by a process referred to as gene targeting). This technique can also be used to remove a gene to produce a null mutation, with a single gene missing (knockout mutation).

In homologous recombination there are two possible alternatives. A sequence can be incorporated in the DNA using sequence replacement or sequence insertion linear vectors. With the sequence replacement technique, the linear replacement vector contains portions homologous to the wild-type gene bracketing the replacement sector. When the homologous segments (genomic and vector segments) are paired, recombination events may occur so that the genomic sequence is replaced by the vector sequence. In the insertion technique, in contrast, the linear vector is engineered so that loci that are adjacent in the genetic map of the genome are at the ends of the vector and separated by the new sequences.

When the homologous sequences are paired to the genomic sequences, the entire vector is incorporated so that part of the DNA remains as a duplicate. Because the recombinations are rare, it becomes necessary either to have selection markers incorporated into the vector (e.g., resistance to drug) or enrichment techniques (e.g., PCR). The identification of correctly modified cells depends on amplifying specifically the recombinant DNA fragments by PCR. This is achieved as follows. The target chromosomal sequence can be represented by A--B and the exogenous DNA as X--Y, where the letters represent DNA ends.

The recombinant fragment will be A--Y. If only the primers for A and those for D are present during the PCR, the primer for A will replicate a sector corresponding to AB and the primer for Y will replicate the XY segment producing single stranded DNA at a linear rate (they can only start replicating at one end). However, the recombinant (A+Y) can synthesize a double stranded DNA with exponential amplification and therefore outpace the production of the other fragments.

The application of homologous recombinations techniques have been used to eliminate genes from specific regions of the brain in mice. These studies used the phage P1-derived Cre/loxP recombination system. The Cre recombinase of the phage catalyzes the recombination between 34 base pairs loxP recognition sequences in vitro or in vivo and the recombination does not require any other protein factors. The loxP sequences can be inserted in the genome of embryonic stem cells by homologous recombination so that they flank one or more exons of a gene of interest.

Mice homozygous for the floxed gene are crossed to a second mouse containing a Cre transgene under control of a cell-type specific transcriptional promoter. In the homozygous progeny containing both the floxed and the Cre transgene, the floxed gene will be removed by the Cre/loxP recombination and only in specific cells. The technique was used to eliminate the NMDAR1 gene in the CA1 pyramidal cells of the hippocampus of mice. In other experiments the NR2B subunit of the NMDA receptor was overexpressed in the forebrain of mice using the CaM-kinase-II promoter which is tissue specific.

An emerging technique allows modifying DNA using introns, referred to as group II introns, which can insert themselves into DNA. After removal from pre-mRNAs, the introns are incorporated into double stranded DNA and transcribed in reverse to form genomic DNA, a reaction requiring a protein with reverse transcriptase, RNA splicing and DNA endonuclease activity and encoded by the introns. The insertion takes place at sites of 14 nucleotides by base pairing interactions.

The site of insertion can therefore be controlled by modifying the introns so that specific genes can be targeted, since introns can be inserted into any DNA with the appropriate target site.

Additional DNA can be introduced in the intron. The intron can be delivered to human cells in culture using plasmids. Gene function can be also manipulated inside cells using synthetic triplex-forming oligonucleotides (TFOs). Intermolecular triplexes can form at the DNA double helix that are purine rich in one strand. The third strand is located in the major groove of the DNA and binds to the purine rich duplex by forming specific hydrogen bonds. The in vitro transcription of RNA polymerase was found to be inhibited by the third RNA strand.

Apparently, triplex formation interferes with transcription-factor binding or the initiation complex. In addition, the triplex can be used to produce site specific damage by attaching covalently to TFOs reagent moeities. The TFO has to be delivered into cells to be effective and a variety of techniques have been devised. TFOs have been shown to produce specific mutations in somatic cells of intact mice. Another technique using hammerhead ribozymes (ribozymes are RNAs with enzymatic activity) that cleave mRNAs at specific sites has been used to block the expression of certain genes.

A dimeric ribozyme construct, called maxizyme has a sensor arm that can recognize target sequences and has been used to inactivate specific mRNAs. In transfection, the genetic information contained in a vector is incorporated in the host's DNA. Transfection of fertilized eggs can lead to the production of stable transgenic organisms that will pass the newly acquired gene from generation to generation. Gene-silencing has been useful in determining the function of specific genes. However, since the function of some of these genes may be essential for the cell's viability, it has become important to generate cell lines or mice that are conditional mutants.

Temperature sensitive mutants have been useful in this respect. One of the ways of producing a conditionally null gene is to keep the cell viable by introducing a transgene whose expression can be controlled experimentally . Recombinant DNA techniques have revolutionized genetics, the study of gene expression, and our understanding of the functioning of protein molecules by making it possible to produce or modify genes, change their genetic environment and change proteins. Used to produce large amounts of proteins and other compounds, recombinant DNA has entirely changed the study of their physiological role and therapeutic use.

USE OF ANTISENSE RNA AND DNA

The function of a gene can be studied by the introduction of an RNA or a single stranded DNA complementary to the mRNA of the target gene into the cytoplasm of cells. This antisense molecule can base pair to the mRNA so that it cannot be translated. There are several ways in which this can be accomplished. Antisense RNA can be synthesized in vitro using RNA polymerase from phages and then microinjected into cells. Similarly, vectors can be constructed to produce high levels of antisense RNA. These can be plasmids containing the target gene in a backward orientation.

It has been possible to produce transgenic mice with these plasmids. Alternatively, single-stranded DNA oligonucleotides can be synthesized with the sequence in the vicinity of the initiation site (the AUG codon). The hybridization of the oligonucleotides to the mRNA will block the initiation of translation. There is evidence of endogenous antisense RNA carrying out regulatory functions physiologically. Double stranded RNA (dsRNA) has been found to silence genes. This silencing is post-transcriptional and takes place in a mechanism in which dsRNA triggers degradation of homologous mRNA in the cytoplasm.

However, at least in plants interactions between homologous DNA and RNA sequences can silence genes by inducing DNA methylation. In addition, RNA-induced chromatin modifications have been observed in plants. Although so far there is no evidence of this latter effect in animals, it would be surprising if the phenomenon were restricted to plants alone. In the nematode Caenorhabditis elegans, specific double stranded mRNA (dsRNA) was found to block selected genes specifically. dsRNA has similar effects in other organisms.

Furthermore, the effect was observed by injecting the dsRNA in the body cavity of the nematode. Apparently, this molecule had no trouble crossing cellular barriers and affects a variety of differentiated cells. In addition, the gene silencing is also transmitted to the the worms progeny. Genetic studies have revealed that the systemic transmission of the RNA interference (RNAi)in C. elegans requires the expression of a number of genes. One of these encodes a protein, SID-1 which has a putative transmembrane domain.

Surprisingly only a few molecules of dsRNA are required per cell, suggesting the presence of a catalytic or amplification mechanism. How the dsRNA acts in Caenorhabditis elegans is still not entirely clear. In experiments using Drosophila cells in culture transfected with specific dsRNA, the cells exhibited loss of function phenotype with a decrease in the corresponding mRNA. Extracts of the transfected cells were found to contain a nuclease which degrades transcripts homologous to the double-stranded RNA.

Apparently, the dsRNA is degraded into short antisense RNA segments, complementary to the targeted mRNA in a process that is ATP dependent. Both strands of the dsRNA are cut into segments 21-23 nucleotides in length, in an RNaseIII-like reaction. The targeted mRNA is cleaved only within the region of identity with the dsRNA at sites 21-23 nucleotides apart. The short segments used for targeting are thought to be incorporated into a ribonuclease complex and to serve as a guide for the degradation.

The endonuclease (Dicer) has helicase activity and contains a dsRNA binding domain and a PAZ domain. The PAZ domain is a region of 110 amino acids involved in the production of siRNA. In a second step, the antisense siRNA, homologous to the suppressed gene, guides another RNase complex, the RNA-induced slicing complex (RISC), to cleave the homologous single stranded mRNAs approximately in the middle of the region paired with siRNA. The use of RNA interferences techniques has been extended by an approach that allows massive screening of the genome of Caenorhabditis elegans whose sequence is known.

Such screens are made possible by feeding the worms with bacteria, each kind expressing double-stranded RNA corresponding to the DNA sequence present in the genome of this nematode. The extensive RNAi library (16,757 RNAs) created in one study is reusable for other studies of gene function in C. elegans. In one study, the mutant phenotypes for 1,722 genes were identified. Genes of similar functions were found to share similar transcription profiles and to be clustered in distinct very large regions of individual chromosomes.

An RNAi inactivation screen of 5,690 Caenorhabditis elegans genes for genes involved in longevity uncovered a mutation in the mitochondrial leucyl-tRNA synthetase gene which was found to impair mitochondrial function and was found responsible for a longer-lifespan. The findings suggest a complex coupling of metabolism and longevity. A similar screen was carried out to disrupt the expression of worm genes involved in normal fat storage using the entire available library. The study identified 305 gene inactivations that cause reduced body fat and 112 gene inactivations that cause increased fat storage.

Many of the nematode genes have mammalian homologues supplying a tool for the study of human function and disease. The possibility of a similar effect of dsRNA in mammalian tissues is intriguing and it has been proposed that the effects ascribed to antisense RNA may be from a contaminating dsRNA.

However, in mammalian cells or in animals intracellular injection of dsRNA is often used as a control in antisense experiments and no comparable results have been reported. It would be interesting to observe what would happen with an extracellular injection in whole tissues or organisms.

In addition to silencing genes in cells, expression constructs can be used to create germline transgenic mice in which a target gene is silenced. In these experiments DNA responsible for expressing short hairpin RNA (shRNA) is introduced into embryonic stem (ES) cells. shRNAs are synthetic molecules mimicking the hairpin secondary structure of microRNAs which cause RNA interference. The ES cells are then injected into blastocysts. The normal function of dsRNA is likely to be the suppression transposons.

Transposons are mobile DNA elements which may contain several genes including those required for transposition and may insert copies of themselves all over the genome. Transposons produce dsRNA. This post-transcriptional gene silencing may be also important in defense from virus invasion or in the regulation of gene expression.

Recognition Techniques

DNA and fragments obtained by restriction endonuclease treatment of DNA can be separated by polyacrylamide or agarose gel electrophoresis. The positions of the bands can be detected by using autoradiography when the DNA has been labelled or alternatively by staining with dyes such as the fluorescent ethidium bromide. Specific base sequences can be recognized by hybridization techniques in which double-stranded DNA is denatured and the resulting single strand is annealed to a probe.

Such a probe is a specific sequence of DNA, usually radiolabelled so that its presence, and hence that of complementary sequences, can be recognized by autoradiography. In the laboratory, the transfer of denatured DNA bands onto nitrocellulose sheets before annealing, proved most practical. This technique is known as Southern blotting. The same technique applied to the recognition of RNA sequences is referred to as Northern blotting. Recognition of a protein by using a specific antibody as a probe is referred to as Western blotting.

A novel approach uses a device capable of identifying individual DNA strands with single-base resolution. The biosensor consists of a DNA oligonucleotide covalently attached to the the lumen of an a-hemolysin (a HL) nanopore. When a voltage is imposed across the pore and single-stranded DNA molecules bind to the attached DNA strand, the ionic current flowing through a nanopore is reduced. The technique is capable of detecting single nucleotide differences in sectors up to 30 nucleotides in length.

A technique that has revolutionized the study of gene expression involves the use of so called gene or DNA chips or DNA microarrays. In this approach, DNA fragments are arrayed at high density on a solid support. This array permits probing the mRNA of cells by hybridization and identifying active genes from whole genomes. DNA microarrays can also be used to study protein-DNA interactions. In practice, the DNA from specific genes can be delivered to glass microscope slides with high speed robotics to form microarrays.

As many as 2,400 yeast open reading frames (supposedly coding for proteins) can be placed on a single slide. The differential expression of the genes is evaluated by hybridization to cDNA labelled with fluorescent nucleotide analogs. A laser scanning device is used to measure the fluorescence, representing the extent of hybridization, and then the relative intensity is expressed in pseudo color. Where two different states are to be compared two sets of fluorescent cDNA probes can be used, seen as red or green.

When the two probes bind to the same DNA, the combination will show as yellow. A clustering alogarithm developed by Eisen et al., allows arranging genes which are apparently coregulated. This provides information on the transcription of genes (for example, underlying cell cycle regulation). These mass-produced microarrays are a new tool for examining in detail gene expression. Studies have been carried out on the human genome with this technology, monitoring the expression of over 1,000 human genes.

Similarly, the temporal program of gene expression accompanying the metabolic shift from fermentation to respiration, was studied using DNA microarrays containing almost all of the genes of Saccharomyces cerevisiae. Another study implicated 800 yeast genes in the yeast's cell cycle. The expression of 650 genes cycling as the consequence of circadian rhythms has been identified with this approach in mammals. DNA microarrays have also been used to differentiate B-cell lymphomas by examining the expression of approximately 17,500 genes heralding possible important medical applications that can be used in diagnosis and treatment.

In this case the DNA microarrays are referred to as Lymphochip and have been also used to define the changes in gene expression that are responsible for the immune system responses or that cause autoimmune diseases. Other recent applications of DNA-microarrays consist in studies of genes expressed in human melanomas and those resulting in metastasis in mice and human melanomas. Four different childhood leukemias were distinguished by hybridization of 63 RNAs samples of four types of the tumors to DNA chips.

A computer based alogarithm (Artificial Neuronal Network, ANN) was set up to recognize and categorize the resulting patterns. The system was able to classify 20 unknown cases of cancer. Reverse-transcription and the polymerase chain reaction (RT-PCR) have been combined in a powerful technique that generates cDNA from mRNA transcripts and amplifies the resulting cDNAs. After gel electrophoresis, the location of the DNAs in the gel can be identified with dyes such as SYBR green or TaqMan. These procedures allow the quantitative evaluation of gene expression in real time, even with very small amounts of RNA.

DNA microarrays can also be used to study protein-DNA interactions. A genetic approach to study the binding of proteins to DNA and RNA is discussed in Section IIB, an extension of the two-hybrid technique of detecting protein-protein binding. This section will discuss biochemical aproaches. Apart from the older ultracentrifugation techniques that can detect a shift in molecular size by changes in sedimentation rates, two techniques have been predominantly used. In the gel retardation technique, the binding of protein to DNA fragments can be recognized by the slowing down of the migration of the fragment in a polyacrylamide gel.

The slowing down results from the increased size of the complex containing the protein and the DNA. This procedure has been very useful in recognizing in vitro binding. Another technique known as footprinting demonstrates binding by virtue of the protective effect of the complexed protein. When the DNA is exposed in vitro to an endonucleases, the sectors which bind the protein are not spliced by the enzymes so that they do not appear when DNA fragments are examined by gel electrophoresis. The missing segments leave an empty space in the gel, the so called footprint. An alternative technique uses dimethylsulfate which cleaves DNA before G residues instead of enzymes.

Again proteins binding to DNA can serve as shield. The chemical technique has the advantage of permitting use in intact cells thereby offering insights on in situ DNA binding. An extension of the microarray techniques produces cells which express a defined DNA. Different cDNAs are placed as microarrays on the slide on which cells are grown. Cells on the spots containing the DNA are transfected. Thousands of cell clusters transfected with a defined DNA provide cells in which a gene product is either overproduced or inhibited. Together with other techniques, the availability of these cell microarrays provide a massive approach to detecting the function of specific genes.

Transfected cell microarrays expressing 192 different cDNAs demonstrated the involvement of proteins in tyrosine kinase signalling, apoptosis and cell adhesion. A powerful method that permits identifying and isolating DNA fragments binding to a known protein is the chromatin immunoprecipitation (CHIP) technique. With this approach the DNA-protein complex is first cross-linked in situ using formaldehyde. Following homogenization of the cells, the preparation is sheared to solubilize the DNA and the complex is precipitated using an antibody for the protein.

After harvesting the complex, the cross-linking is reversed (e.g., with heat). Most frequently the DNA is subjected to PCR amplification to facilitate identification. Nucleic acid probes can be used in the techniques of in situ hybridization. The DNA can be exposed temporarily to a very alkaline pH to separate the strands. The specific sites can then be recognized with an appropriate probe that will hybridize to the site under study. Modified nucleotides can be used to provide either a target to antibodies or to some other molecule that can be recognized such as biotin that can be recognized by its binding to Texas red-avidin.

The techniques in which fluorescent probes are used for in in situ hybridization is known as fluorescence in situ hybridization (FISH). In situ localization of DNA may be also carried out using sequence-specific DNA-binding proteins. The proteins can be identified by immunological techniques either at the light or EM level. Similarly a green fluorescent protein-lac repressor fusion protein was used using the reporter gene technique.

A piece of DNA coding for a protein that is easily recognizable is attached to the gene under study. Favorite reporter genes have been the lacZ gene of E. coli which codes for b-galalactosidase or the gene for chloramphenicol acetyl transferase of E. coli or, alternatively, the firefly gene coding for luciferase. Most recently green fluorescent protein (GFP) has gained favor.

Recognition and Separation of Proteins

Proteins are the basic units of structure and function of cells. The recognition and localization of proteins are therefore crucial to the understanding of cell organization. In addition, the ultimate measure of gene expression is the production and covalent modification of proteins. mRNA abundance, detectable using the DNA microarrays does not necessarily correlate with the production of proteins.

Use of antibodies

Injection of an antigen into an animal elicits the production of antibodies originating from different antibody-producing cells, each responding to a different epitope (i.e. region of the antigen molecule). However, a single antibody producing cell and its progeny (i.e., a clone) will synthesize only a monoclonal antibody, specific for one or a few closely related epitopes. Techniques that take advantage of these properties have been developed using myeloma cells. Multiple myeloma is a malignancy of antibody-producing cells.

Many cells (a clone) can be generated from the proliferation of a single myeloma cell. In practice, antibody-producing cells are fused to myeloma cells. Lymphocytes and plasma cells are obtained from the spleen of a mouse that has been immunized against an antigen. They are then fused to myeloma cells by special procedures. The hybrid cells (hybridoma cells) are immortal and continue to proliferate in vitro, in contrast to normal cells such as nonfused cells from the spleen of the immunized mouse.

The hybrid cells can be selected from the unfused malignant cells if the myeloma cells used in the fusion had a biochemical defect. Only the fused cells containing the wild-type genome of the spleen cells can survive. In the standard techniques, the defect is absence of the enzyme hypoxanthine-guanine phosphoribosyltransferase (HGPRT). The cells generated from one of these fused cells will produce a clone of identical cells producing the monoclonal antibody. Monoclonal antibodies can be used to recognize specific groups in proteins and, when incorporated in affinity columns, to aid in the isolation of a specific protein from a mixture.

Antibodies can also be produced by other means. These procedures are simpler and have proven very practical. When injected into mice, PCR-amplified DNA open-reading frames (ORFs) noncovalently linked to a eukaryotic promoter and terminator, produce antibodies specific for the encoded protein. Monoclonal antiodies can also be produced by display technology. In the immune system the rearrangement of V gene segments produces the repertoire of antibodies, each capable of interacting with one antigen.

These gene segments can be amplified using PCR. When these genes are cloned in phages the variable antibody domains are displayed at the surface, so that they can be selected by antigen coated plates or columns containing the antigen. In immunoprecipitation antibodies cross link antigen molecules so that they form large insoluble complexes and can be readily isolated. In studying the location of specific components of cells, either heterogeneous or monoclonal antibodies are conjugated to markers.

For light microscopy, a fluorescent dye is frequently used. Alternatively, an antibody to a specific protein is used and the fluorescent dye is attached to an antibody that recognizes the constant region of the primary antibody. Submicroscopic markers, such as ferritin or colloidal gold, are used with electron microscopy. Immunological techniques have provided valuable information about cell structure and function.

Immunological methods can be used to measure the amount of a specific protein present in an extract. One technique has combined PCR with conventional immuno-detection methods (Immuno-PCR). More recently, another technique has combined immunoprecipitation with T7 RNA polymerase amplification (IDAT). In Immuno-PCR, a linker molecule with binding affinity for DNA and an antibody (Ab) is used to attach a marker DNA molecule specifically to an antigen-Ab complex to form a specific antigen-Ab-DNA conjugate.

After precipitation of the complex, the attached marker DNA can be amplified by PCR and eventually measured. This technique has a sensitivity 105 higher than the more conventional antigen detection systems. IDAT is 109 more sensitive than conventional immunological methods. It consists of attaching double stranded oligonucleotides containing the T7 promoter to antibodies against the antigen being studied. The RNA is amplified with T7 RNA polymerase using radioactive nucleotides and the radioactive product can be used to quantify the procedure. Immuno-PCR and IDAT can be used to monitor proteins, lipids, and metabolites and in the case of IDAT, at the single-cell level.

Fluorescent proteins

The use of fluorescent probes including fluorescent proteins and low molecular weight probes has been throughly discussed recently. The small molecules form a fluorescent covalent complex with any intracellular protein to which a reactive group has been genetically fused. Some of the most useful techniques will be discussed in this section. The jelly fish Aequora victoria emits blue light. It contains two photoactive proteins. The protein aequorin is bioluminescent and emits blue light when complexed to Ca^{2+} and the fluorescence is proportional to the concentration of Ca^{2+}.

Aequorin can be microinjected into cells or introduced into cells by cDNA transfection. An additional tool is provided by the targeting of aequorin to specific compartment of the cell by introducing in the cDNA codes for targeting the protein to specific localities. A second protein, the green fluorescent protein (GFP) absorbs blue light and emits green light. In the intact organism the two, aequorin and GFP, act together to produce the green light of the jelly fish. GFP has been used in a variety of experiments because it is stable and easily visualized microscopically.

Chimeric DNA coding for a protein and GFP serves as a useful marker to trace the fate of newly synthesized proteins. It has been used, for example, to examine the fate of proteins leaving the ER. The major disadvantage of the approach is that GFP can interfere with the function of its fusion partner. Fortunately, this may be the exception rather than the rule. GFP-fusion proteins containing targeting information (in the form of an amino acid domain such as a signal sequence) can be produced by manipulation of cDNA. In this way, the GFP can be targeted to specific locations.

This approach can identify the location of proteins or structures in a living cell without further manipulation. The necessity of producing a new cDNA makes it relatively easy to subject it to site-directed mutagenesis. Mutants of GFP are available that differ in excitation and emission peaks so that several proteins can be tagged in the same cell. Many fluorescent indicators derived from GFP are currently used. Some respond to pH, others to Ca^{2+} and Cl^{-} concentrations. Any of these can be targeted to specific intracellular compartments.

GFP attached to repressor molecules has permitted identifying the position of operators in the study of gene expression in living cells. Like any other fluorescent probe, GFP and GFP mutants can be used in fluorescent recovery after photobleaching (FRAP) studies. One added feature is that the bleaching is reversible in GFP mutants with two excitation peaks.

When the bleaching is carried out at the longer wavelength, illumination at the shorter wavelength reestablished the fluorescence.

Techniques capable of detecting biological ligands such as Ca^{2+} or cAMP take advantage of the fluorescence resonance energy transfer (FRET)between different mutants of GFP. FRET can take place between two chimerically connected GTP variant proteins, from shorter to longer wavelength (such as the transfer of energy from blue emitting protein to green or yellow emitting mutant proteins). The excitation of the donor nonradiatively excites the recipient and the latter emits a fluorescence at a characteristic wavelength.

This transmission requires that the two variants of GTP be appropriately positioned. In principle, the GTP-protein pairs can be bridged by protein domains that change in conformation when they bind a biological ligand. These chimeric proteins can be used as indicators of the ligand. They can be introduced into cells via transfection of the modified gene. The chimera can also contain targeting sequences that direct them to specific locations. For example, calmodulin or calmodulin domains were connected to GTP-protein pairs in such a way that the distance or geometry between the pairs was altered by Ca^{2+} binding.

The chimeric proteins could therefore be used as calcium indicators. The introduction of a targeting sequence directed the protein to the lumen of the endoplasmic reticulum. Many more applications of GFP technology have been developed in recent years. Adding the appropriate targeting signal to GFP, has made it possible to direct GFP to specific intracellular compartments. The fate of an RNA transcript has been followed by constructing a plasmid containing the DNA coding for an RNA-binding protein fused to GFP.

Time dependent events can be followed in cells using a mutant (E5) of a red fluorescent protein (drFP583) isolated from corals (Anthozoa). E5 was found to change in fluorescence from green to red in a time dependent manner. Transgenes in which E5 was controlled by the promoter of heat shock protein or the Xenopus Otx-2 promoter were microinjected into Caenorhabditis elegans and Xenopus laevis embryo cells respectively. These experiments demonstrated that E5 could be used to monitor activation and down-regulation of target promoters.

Other protein probes

In addition to the DNA and antibody probes already discussed, other techniques are available. Affinity probes can be used in the study of binding sites to specific ligands in enzymes or other proteins. Affinity labels resemble the natural ligand, but they bind covalently. Photoaffinity labels can be used in a similar manner. They are generally analogues of the natural ligand, usually azido compounds that are stable in the absence of light.

When photolyzed by exposure to light of the appropriate wavelength, they form highly reactive groups that bind to proteins. The recognition of special binding domains in proteins permits the binding of probe molecules (e.g., proteins attached to the GFP) to proteins. The specificity of binding depends on the selection of the appropriate probe molecule. This technique has been used to pinpoint the location of certain phosphoinositides in cells using proteins containing the PH domain.

Separation techniques

Many methods have been used to isolate, identify, and purify proteins. Some of these stand out for their usefulness or frequency of use. Integral proteins, embedded in membranes require the use of detergents for their extraction. The detergents bind to the hydrophobic portions of the lipids through their hydrocarbon chains and interact with water through their polar groups. The use of amphipols promises a new era in the study of membrane proteins.

Amphipols are polymers with alternating hydrophilic and hydrophobic side chains. They can solubilize integral membrane proteins by surrounding the hydrophobic protein domains with a polar exterior. In aqueous solutions, the amphipols bound integral proteins studied are in their native state (bacteriorhodopsin, a bacterial photosynthetic reaction center, cytochrome b6f, and porin). In at least the case of diacylglycerol kinase, an integral protein, amphipols have been shown to support enzyme activity normally requiring polar lipids such as cardiolipin.

Antibodies can be used to precipitate (immunoprecipitate) specific proteins. On incubation of a mixture with a specific antibody, the antigen is bound. Multivalent antigens and multivalent antibodies form large complexes, which precipitate. With the exception of monoclonal antibodies, serum contains a variety of antibodies for different determinants in the antigen molecule. Precipitation can also be produced by adding Staphylococcus A cells. These cells contain at their surface, a protein that binds the constant region of most antibodies, producing very large aggregates.

Obviously, specific antibodies cannot be produced unless a pure protein is available first to serve as a specific antigen. The techniques used for the purification of proteins are varied. Some of the classical approaches depend on the solubility of proteins under a variety of conditions and concentrations of salts (such as ammonium sulfate) or binding of proteins to columns. In ion-exchange chromatography, proteins are separated according to their charge. An ion exchanger, either an organic cation or anion, is bound to supportive material.

The proteins separate out by binding differentially to the ion exchanger. Since the charge of the protein depends on its degree of ionization, the protein charge, and hence the degree of separation, depends on pH. Many other techniques are based on separation of proteins by size. Among many others, they include gel filtration, density gradient centrifugation, and sodium dodecyl sulfate-polyacryl-amide gel electrophoresis (SDS-PAGE). In gel filtration, separation of proteins depends on passage of the soluble mixture through a column containing hydrated carbohydrate beads.

The smaller the protein, the larger the portion of the water in which it will be distributed, so the larger molecules will migrate faster through the gel. The rate of passage of various protein molecules of known size allows a molecular weight calibration that can be used to size other proteins. In electrophoresis, proteins are separated in an electric field because of their charge. When a polyacrylamide gel is used, it acts as a restrictive matrix that also separates proteins by their size. In SDS-PAGE, the proteins are denatured and reduced so that the component polypeptides are separated.

Furthermore, they are coated with the detergent SDS, which is negatively charged. Ideally, this coating is evenly distributed, so that the proteins will separate electrophoretically by size only. The protein bands are stained or otherwise made visible. Affinity chromatography is potentially one of the most powerful techniques. In this technique a specific ligand (e.g., a cofactor, substrate, or antibody) is attached to a matrix. When an extract containing proteins is passed through the column, only the targeted protein remains in the column and can be subsequently eluted (e.g., by free ligand).

In high-performance liquid chromatography (HPLC) the separation uses narrow and long columns of tightly packed glass or plastic beads coated with a thin layer of stationary phase. The fluid phase is forced through at high pressures. The procedure is very rapid. However, the size of the samples is necessarily limited. Proteins that specifically bind to other proteins can be identified by an ingenuous approach referred to as interaction cloning. With this technique, the cDNA of a known protein is modified by adding the coding sequence for the phosphorylation site of heart muscle kinase.

The composite protein is then expressed in E. Coli and after purification and labelling with [^{32}P], it serves as a probe for a λgt11 cDNA expression library from eukaryotic cells under study. The E.coli plaques containing the unknown protein binding the radioactive probe can then be detected and the λgt11-cDNA isolated and cloned. In the study of Blanar and Rutter, the purification of the probe was carried out by immunoprecipitation with a monoclonal antibody to an epitope added to the probe protein.

Gene expression and identification of proteins

The identification of proteins expressed under different conditions (for example, during development or in disease) using microarrays holds considerable promise). This approach is similar to the one using DNA microarrays. A study of protein binding using protein microarrays is discussed later. In general, the possibility of creating protein microchips has been beset by several difficulties. Whereas it is possible to amplify mRNA by the use of reverse-transcriptase polymerase reaction, no such amplification device is available for proteins.

Furthermore, the proteins must be immobilized in their native conformation with their active sites exposed. In addition, the proteins are extremely heterogenous so that it is difficult to develop a surface to bind them all. The use of antibody microarrays to detect the presence of specific proteins is probably one of the most general techniques available. In another approach, proteins with differing surface chemistries are immobilized in arrays taking advantage of these differences. The bound proteins are then examined with time-of-flight mass spectrometry.

FUNCTION OF PROTEINS

In Situ Protein Inactivation

The technique of chromatophore-assisted laser inactivation

(CALI) allows for the selective inactivation of proteins. CALI requires conjugating a dye (generally malachite green) to a non-blocking antibody. Laser light at 620 nm (for this particular dye) is used to inactivate the protein by generating short lived hydroxyl radicals (psec). Generally, neighboring proteins remain unaffected. In contrast to proteins at the cell surface, intracellular proteins can only be studied by microinjecting the dye-conjugated antibody.

The technique is not without its problems. The distance between dye and target protein may be too great for effective inactivation, or the protein may be particularly insensitive to hydroxyl radicals. Furthermore, in the case of an abundant protein (e.g., tubulin or actin), the possibility of inactivating all of the molecules present is remote. However, recovery by diffusion from other parts of the cell or de novo synthesis may be relatively fast. Therefore, a negative result does not necessarily indicate a lack of involvement in a function. The effect might also be indirect. In short, data obtained with this approach (as with any other) must be carefully evaluated.

Dominant Inhibitory Proteins

Dominant-inhibitory mutant proteins suppress the activity of the wild-type proteins. Therefore the expression of an inhibitory mutant can be used to delineate the biochemical function of the normal protein. The mutant gene can be introduced by transfection. This approach has been used extensively in the study of the function of certain proteins such as the Ras-family GTPases. In general, as many as 6,000 studies of signaling molecules such as kinases, phosphatases and transcription factors have used dominant-inhibitory proteins.

A more direct approach involves the removal of a specific protein. The protein is targeted for degradation by the proteasomes. In the proteasomal degradation pathway, substrate specificity of the E_3-complex (such as the SCF complex) is determined by F-box proteins which serve as receptors for the substrates. Therefore, the introduction of a specific interaction domain in the F-box proteins allows binding of the targeted protein to E_3-complex and hence its degradation.

Binding properties

The determination of the rate and binding of components, particularly proteins, has become increasingly important. Since protein-protein interactions play such an important role in function, determining these interaction may provide a basis for mapping their function. The role of protein-protein interactions are many. Enzymatic reactions have frequently been found to involve complexes, and hence binding of proteins to other proteins.

For example, RNA polymerase II can function as a complex of 12 subunits. However, further studies found that it is comprised of 55 subunits to reconstitute the native transcriptional activity. Regulatory signals appear to involve protein complexes organized around scaffolding proteins. In addition, where enzyme and substrate are both proteins, the two may be associated in a stable manner. Generally, the studies of binding of proteins to other proteins, have used biochemical techniques. Three new approaches, genetic and biophysical, have been used.

Biophysical techniques

Biophysical techniques have proven very useful. The use of surface plasmon resonance (SPR) has become common (commercialized by Pharmacia, Uppsala, Sweden, as BIAcore). The technique is a real time optical detection system. In this technique, one of the interactants is immobilized on a dextran-coated gold surface. The second interactant is then injected across the surface and the interaction is monitored optically continuously.

The technique is based on the fact that when monochromatic light is polarized and incident on a thin metal film-liquid interface, a component of the incident light momentum (the evanescent wave) penetrates a distance of about one wavelength into the less dense medium. The evanescent wave interacts with free oscillating electrons (the plasmons) in the metal surface. When plasmon resonance occurs, energy from the incident light is lost to the metal film, decreasing the reflected light intensity.

The resonance occurs only at a precise angle of incident light and this angle depends on the refractive index of the medium. The refractive index is altered by the interaction of macromolecules and detected as a change in resonance angle. The fluorescence of a single molecule bound to a surface component can be detected microscopically using total internal reflection. The technique has been applied using a fluorescent ATP analog bound to myosin. The ATP analog bound to a single myosin immobilized on a microscope slide is excited by a laser beam.

The method depends on the total internal reflection of the fluorescence between the interface of the slide and the solution. The light penetrates the solution for a fraction of a wavelength, exciting primarily a molecule in contact with the surface rather than in the bulk solution. A stationary fluorescent spot indicates binding of ATP (or ADP) to the attached myosin. Disappearance of the spot indicates dissociation from myosin of either ATP or the ADP produced by hydrolysis. A similar total internal reflection technique was used with a single fluorescently labelled myosin subfragment-1 attached to a mechanical probe.

This approach was also used to follow the exocytosis of synaptic vesicles labeled with a fluorescent lipid in goldfish retinal bipolar neurons. The synaptic terminal attaches tightly to a coverslip. A thin layer of cytosol beneath the plasma membrane is illuminated selectively and then fluorescence emission reaching the microscope objective is collected. The light reflected from the interface between glass and terminal produces a field that excites any fluorescent molecule within a 100 nm of the site. The brightness of a fluorescently labeled vesicle increases as it approaches this site. At the fusion, the fluorescence is incorporated into the plasma membrane.

Microarrays and display techniques

The binding of proteins to other proteins or molecules has begun to be studied by microarray technology that produces robotically small spots of immobilized proteins on glass slides treated with an aldehyde containing silane reagent. The aldehydes react with primary amines in the proteins. The proteins attach in various orientations so that the reactive groups of some of the molecules are not shielded from reagents or other proteins. With this technique, one protein present as 1600 spots per cm^2 can be tested for binding to other proteins or other chemicals labeled with a fluorescence probe.

A similar technique was used to examine the interaction of proteins representing 5800 open reading frames of yeast with other proteins and phospholipids. Many new calmodulin- and phospholipid-interacting proteins were identified. Many interacting proteins contain several proline residues. Other proteins contain a proline residue in their critical sequences. The interactions between two proteins generally involve short segments, for example, in antibody-antigen reactions the epitope generally is composed of only 4 to 7 amino acid residues.

Similar conclusions were reached for Src homology (SH) 2 and 3 domains, phosphotyrosine binding domains (PTB), postsynaptic density/disc-large/ZO1 (PDZ), WW domains, Eps15 homology (EH) domains and 14-3-3 proteins where they generally recognize sequences of 3 to 9 amino acids. SH2, SH3 and the Pleckstrin homology domain (PH) are found in all eukaryotic organisms, except for SH2 that is not found in yeast. Display technologies have been very valuable in screening for protein-protein interactions.

Although, like the microarray techniques, they are in vitro techniques raising questions about their relevance, they have the advantage over the two-hybrid techniques described below in allowing testing a greater number of interacting proteins and are not limited to interactions occurring in the yeast nucleus. The phage-display techniques insert a DNA fragment coding for a particular protein in a portion of the phage-DNA coding for coat proteins. Therefore, after the phage reproduces inside the E. coli host, the protein in question will be displayed at the surface of the phage and therefore will be exposed to protein ligands that can be isolated readily.

In the in vitro ribosome display procedure, the cDNA for a protein is amplified and then a T7 promoter, ribosome binding site and stem-loops are introduced . After translation, the process is stopped by cooling and the RNA-ribosome-protein complexes stabilized. Then the protein can be selected by binding to interacting protein. After isolation and dissociation of the mRNA, the RNA is reverse transcribed and the DNA amplified by PCR.

In the mRNA display technique, covalent attachment between an mRNA and the protein that it encodes takes place by in vitro translation of synthetic mRNAs that carry puromycin, a peptidyl acceptor antibiotic at their 3' end. The joining of the two molecules allows a specific mRNA to be selected from a mixture according to the properties of the protein. This mRNA could then be amplified by reverse transcription followed by PCR.

Genetic approaches

One of the genetic approaches, the so called two-hybrids technique, takes advantage of the fact that gene activators have two independent domains. One domain binds DNA (DB) and the other is responsible for the activation (AD). If hybrid proteins are produced, one protein (e.g., Y) attached to the activating segment(AD-Y) and the other (e.g., X) to the DNA-binding domain (DB-X), the two will be able to activate the gene only if they bind to each other. In the original study, the domains of the GAL4 protein of Saccharomyces cerevisiae were used and the hybrid proteins were produced by transferring vectors containing the appropriate coding DNAs into a yeast strain lacking GAL4.

The yeast expressing GAL4 were selected and the plasmids isolated. The use of growth selection markers (e.g. Leu1 and HIS3) rather than Gal4p, has allowed the use of the powerful growth selection. The two hybrid technique can be modified to recognize RNA-protein binding and DNA-protein binding . An extension of the two-hybrid protocol permits identifying DNA encoding proteins capable of binding to DB-X from DNA libraries. DB-X was referred to as "bait". Plasmids are constructed to encode two hybrid proteins. One hybrid consists of the DB domain of the yeast protein Gal4p fused to the test protein.

The other hybrid consists of the Gal4p activation domain fused to protein sequences encoded by a library of genomic DNA fragments. Binding of the test protein and a protein encoded by one of the library plasmids leads to transcriptional activation of a reporter gene containing a binding site for GAL4. Use of the appropriate cDNA libraries in the activation domain plasmid could extend this technique to any system. This approach was used to identify mammalian proteins that bind to the proteins Jun or Fos and to isolate cDNAs that encode a protein that is recruited to the c-Fos serum response element (SRE).

Two-hybrid analyses in yeast can also be carried out by introducing different "baits" and "preys" in the same diploid cells by mating. In an attempt to characterize the functions and the relationships of proteins, a massive yeast two-hybrid screen was undertaken. 6,000 yeast colonies were screened. Each expressed a different "prey" molecule fused to an activation domain and strains expressing 192 different "baits" were mated to each member of the "prey" colonies.

Alternatively, the "prey" cells were mixed and mated to each different "bait" colony. The library method has the disadvantage that competition between proteins can mask interactions. 957 probable interactions involving 1004 proteins were detected. The function of some proteins could be deduced from the partner they were interacting with. New interactions were discovered, particularly in relation to the cell cycle. For example, the cyclin dependent kinase, Cks1, was found to interact with three different B-cyclins, so that Cks1 may have a role in activating Cdc28 kinase and initiate the cell cycle.

Cdc28 (Cdc2) is the cyclin-dependent kinase of and Cks1/ Suc1 is a regulatory protein associated with Cd28. In addition to these, the study uncovered a multitude of other interactions including those of prcteins involved in meiotic recombination. A modification of the two-hybrid technique involves the use of counterselection markers and has been called the reverse two-hybrid system. The expression of these markers are lethal under certain conditions. For these cases, the selection is for the dissociation rather than binding.

The technique can be used to screen for mutations that block specific associations and for low molecular weight regulators that block the association of proteins. The two-hybrid system is extremely powerful. Two-hybrid screening using a library of random genomic fragments was able to identify contacts between individual polypeptides of polymerase III and its interaction with TFIIIC. Similarly, a large scale two-hybrid analysis of 27 proteins involved in vulval development of Caenorhabditis elegans produced a protein interaction map of this system.

The two-hybrid system provides an answer for in vivo interactions and allows the assay of a multitude of of coding sequences rather simply. It should be noted, however, that false positive or false negative results do take place and the two hybrid technique requires additional validations. Although the yeast two-hybrid method has been extremely productive in detecting interactions between protein molecules, the proteins under study must be present in the nucleus. Consequently, the method is unsuited for the study integral proteins which are present in membranes.

Integral proteins carry out fundamental functions such as solute transport, signaling, channels, electron transport, etc. Furthermore, approximately 30 % of all proteins coded by the genome are thought to be integral proteins . It is therefore important to be able to determine which proteins interact with integral proteins. The ubiquitin-based split-protein sensor (USPS) method is not limited to proteins present in the nucleus. Protein constructs containing ubiquitin are cleaved in vivo by ubiquitin-specific proteases (UBPs) which recognize the folded conformation of ubiquitin.

Independently from UBPs, ubiquitin itself can be split into amino-terminal (Nub) and carboxy-terminal fragments (Cub). The two fragments remain attached without a covalent bond and the dimer remains functional. In USPS, a reporter protein attached to Cub will be cleaved by UBP when the two ubiquitin fragments assemble. The technique uses a mutationally altered Nub unable to attach to Cub. However, if the two ubiquitin fragments are fused to interacting proteins, Nub and Cub will be in close contact, activating the release of the reporter protein from Cub by the action of the UBPs.

The reporter protein can be detected in a variety of ways (e.g., immunoprecipitation and Western blot analysis of the cleaved reporter protein). Alternatively, the reporter protein can be an enzyme, e.g., Ura3p which would enable yeast cells to form colonies on media lacking uracil or a transcription factor, which would activate reporter genes. USPS provides a generally applicable assay for in vivo protein interactions which makes it possible to monitor a protein-protein interaction as a function of time, at the natural sites of this interaction in a living cell.

The reverse Ras recruitment system (RRS) takes advantage of the properties of the Ras GTPase system in yeast. Ras is present in the plasma membrane. The yeast Ras-guanyl-exchange factor Cdc25 present in the membrane interacts with Ras and stimulates its guanyl nucleotide exchange and ultimately produces cell growth. A temperature sensitive mutant (cdc25-2) grows at 23o but not at 36o. In the RRS method, a membrane protein is expressed in the membrane of yeast (X). Its possible interaction partner (Y) is fused to a mammalian Ras that is cytoplasmic but binds to the membrane and is able to function with Cdc25-2 .

An interaction between X and Y and hence Cdc25-2 and the mammalian Ras permits growth of the cdc25-2 mutant at the non-permissive temperature. In a screening method to detect proteins interacting with X, a cDNA library is expressed as a fusion with the mammalian Ras. The use of this technique is limited by the fact that the fusion of membrane-associated proteins to mammalian Ras will also lead to growth even if there is no interaction between these proteins. Another technique based on the yeast two-hybrid method uses inactivation of G-protein signaling.

The G-protein constituted by $\alpha\beta$ and γ subunits is attached to a receptor. In the G-protein technique, an integral protein under study (X) interacts with a soluble protein (Y) attached to the γ subunit of the G-protein. The interaction between the integral protein and the soluble protein would disrupt the G-protein by sequestering the G β subunit which remains attached to the γ subunit bound to Y. Colonies in which interactions occurred can be detected by examining the sensitivity of the yeast cells to a factor. The cells should stop growing.

Other macromolecular interactions where proteins bind to DNA or RNA are as important as protein-protein interactions. In fact, DNA-protein interactions are responsible for the control of gene expression. A number of variations of the original two-hybrid approach have been developed for the study of nucleic acid-protein interactions.

Chromatin immunoprecipitation and DNA microarrays have been used in the study of the interactions between DNA and proteins. In immunoprecipitation formaldehyde is used to crosslink the proteins to the DNA. Then the crosslinked material is precipitated immunologically. The DNA microarray technique follows the same general concept used for studying DNA-RNA binding.

Sequencing of Proteins

Techniques are available for sequencing proteins; in fact, the methodology has been automated. In practice it is easier to sequence DNA because of the capacity to multiply the DNA either through cloning or PCR, so that the amino acid sequence is most frequently deduced from the nucleotide sequence. The methodology has been standardized to the extent that it is possible to commercially contract for deriving sequencing information from purified compounds.

Conversely, it is possible to synthesize peptides or polynucleotides. A known sequence of a protein can be compared with others by the use of software and databases presently available. Similarly, it is possible to search for particular motifs in proteins (or nucleic acids) through computer searches. One of the interesting applications of this approach is the deduction of function by identifying amino acid sequence domains that are in common with proteins of known function.

The genome sequence of some organisms is now known and the time is approaching when the complete human genomic sequence will be known. The advent of DNA microarrays or chips permits the study of the expression of most if not all of the genes of an organism. However, very little is known about the function of the gene products. Mass spectrometry and sequence database searching have been very useful tools in these studies.

Recent advances in protein characterization via mass spectrometry allow linking these DNA sequences to the proteins in functional complexes as had never been possible before. For example, use of the DNA and protein data bases in conjunction with mass spectrometry have allowed the characterization of the entire spliceosome complex containing as many as 46 distinct proteins. Mass spectrometry is a technique capable of weighing individual molecules by converting them into ions and subsequently measuring their trajectories in response to electric or magnetic fields.

Until the advent of electronspray techniques, mass spectroscopy was difficult to apply to biological molecules because of their fragility. The spray techniques make use of high electric fields to desorb ions from small charged droplets of solution into gas. After partial protease digestion, component peptides are individually processed. Usually, each is randomly cleaved at the peptide bonds by collision with a gas stream, usually He, forming smaller peptides.

In mass spectrographs of these smaller pieces, the fragments will be in order of mass (expressed as mass/z, where z is the charge of the molecule, usually H^+). Analysis of the differences in mass of the different peptides permits deducing the amino acid sequence. The methodology for the identification of peptides from mass spectrographic data has been worked out. In tandem mass spectroscopy two consecutive mass analyzers are used. The first generates and selects the ion (polypeptide + H^+).

This is followed by the fragmentation in a separate chamber and the analysis of the products by a second mass spectrometer. Each peak is generated by the selected precursor. Refining of these techniques allows sequencing proteins from as little as 5 ng samples obtained from polyacrylamide gels. Electrospray-tandem mass spectroscopy has been used in conjunction with microcapillary liquid chromatography and chemical reagents that have been called isotope-coded affinity tags (ICAT) to measure the differences in protein expression in Saccharomyces cerevisiae under different metabolic conditions.

The ICAT reagent has specificity toward sulfhydril reagents and contains a biotin tag. With this technique, a reduced protein sample is tagged with light reagent and another sample, representing a different metabolic state is tagged with heavy reagent. After the two samples are mixed they are enzymatically cleaved to produce peptides. The peptides are then isolated by avidin chromatography. The isolated peptides are then separated and analyzed by microcapillary liquid chromatography and electrospray-tandem spectroscopy.

The ratio of heavy versus light component provides a measure of relative expression of a protein. The various approaches have provided a wealth of information about the organization of cells. The significance of the crowding of macromolecules in the various aqueous compartments of the cell has become grounds for debate. It has been suggested that the crowding of macromolecules may result in the hindering of free exchanges and the alteration of physical properties such as changes in binding coefficients. The diffusion of the various molecules will depend on solute properties as well as the composition and structure of the cellular compartments.

Presently the measurement of diffusion can be carried out using fluorophore labelling and fluorescence recovery after photobleaching (FRAP). The labelled molecules can be introduced by microinjection or transfection. The movement of molecules is slowed down significantly by binding to structures. The slowing down of the diffusion of macromolecules or even small molecules in the cytoplasm is considerable, particularly in the case of very large molecules. However, in most cases it is not likely to be sufficiently slowed down to play a significant biological role.

The diffusion of glycolytic enzymes in the cytoplasm is much slower than that of molecules of comparable size, suggesting significant binding to structures. In addition, the possible dependence of the binding on metabolic states is tantalizing. For orientation, we will first examine a cell as viewed with phase contrast microscopy. Then proceed to a view of the same cells by EM. A photomicrograph of a spread-out living fibroblast obtained with a phase-contrast microscope is shown in Figure. 2.

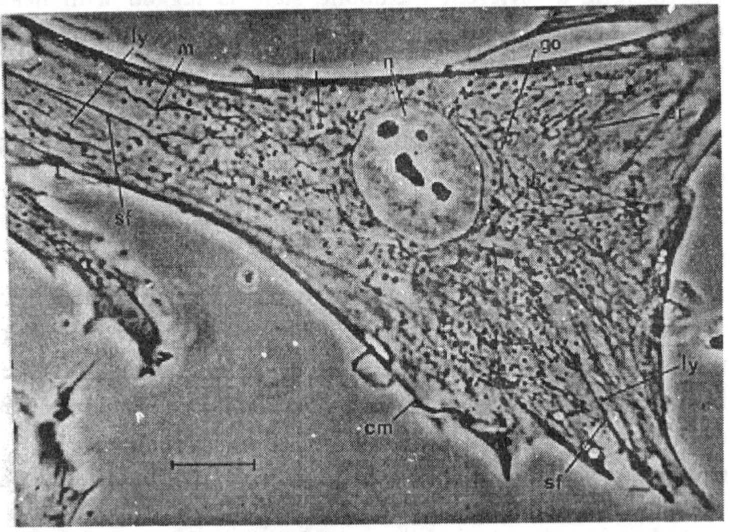

Figure 2: Phase-contrast photomicrograph of living cultured rat embryo cell with its nucleus (n) surrounded by Golgi apparatus (go), endoplasmic reticulum (er), mitochondria (m), lipoprotein inclusions (l), and small vacuoles, which are probably lysosomes (ly).

The technique shows a good deal of cellular structure but without much detail. The nucleus is clearly visible, and lamellar elements that are known to be associated with the Golgi apparatus can also be distinguished. There are various cell inclusions or organelles, filamentous mitochondria, spherical lysosomes, and fibers called stress fibers present in focal contacts. Many variations on this basic plan occur. Some cells have specialized surfaces and specialized ends. The luminal side corresponds to the surface at which the secretory products are

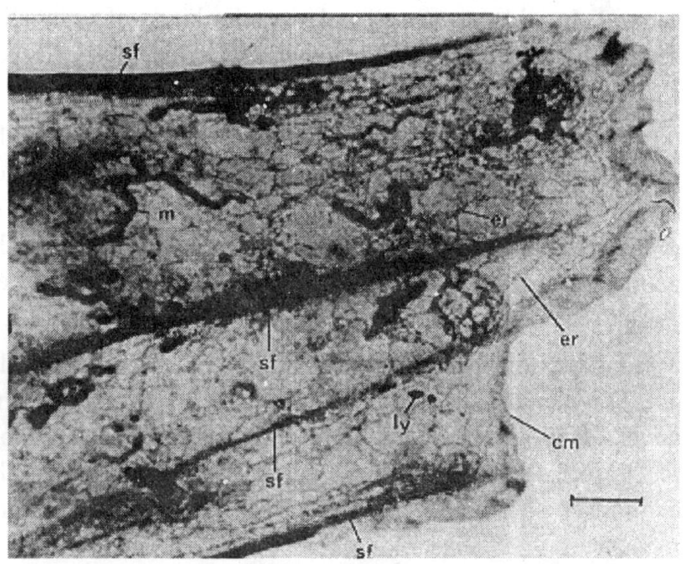

Figure 3: Electron micrograph of projecting process of an unsectioned glutaraldehydeosmium-fixed cultured rat embryo cell showing, within the cell margin (cm), filamentous mitochondria (m), lysosomes (ly), the network of the endoplasmic reticulum (er), and dense linear structures, the stress fibers (sf).

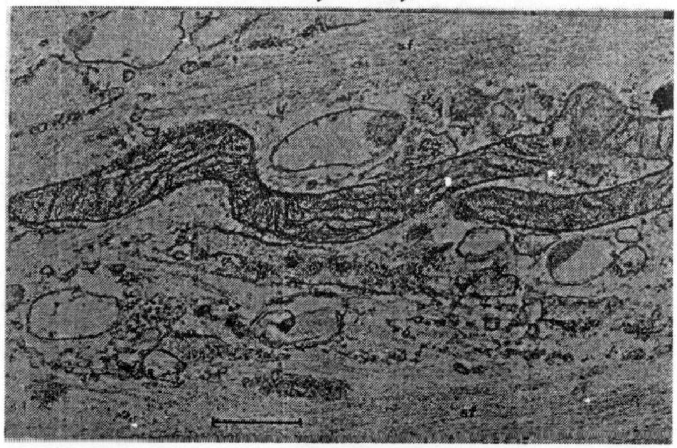

Figure 4: Section through cultured rat embryo cell showing cytoplasmic structures in greater detail

discharged. In addition to cell polarity, the kinds of organelles present or their relative proportions may differ from one cell type to another or from one organism to another.

In figure 2, In several places around the cell margins, the cell membrane (cm) is engaged in undulating movements, which at the lower cell process are associated with the formation of pinocytotic vacuoles. A small number of phase dark lines, the intracytoplasmic stress fibers (sf), are seen within the cell processes (bar corresponds to 10 mm).

Transmission electron micrographs of spread-out cells (Figure.3) without sectioning add little to the phase micrographs. However, sectioning and magnification provide considerable detail (Figure.4). The vacuoles are clearly vesicles, and the mitochondria show both an inner and an outer membrane component.

In figure 4, Centrally, a branching mitochondrion is surrounded by profiles of the endoplasmic reticulum, some of which are studded with ribonucleoprotein particles and contain a somewhat granular material. Flanking this cluster of organelles are portions of sectioned stress fibers (sf), which appear as bundles of near-parallel filaments of about 7.5 nm diameter (bar corresponds to 1 mm).

Chapter 3

Cell Culturing Techniques

Hundreds of commercial micropropagation laboratories worldwide are currently multiplying large number of clones of desired varieties and local flora. Apart from the rapid propagation advantage, this technology is being used to generate disease-free planting material, and has been developed and applied to a wide range of crops, and forest and fruit trees. However, in many cases, the cost of micropropagule production precludes the adoption of the technology for large-scale commercial propagation.

Low-cost tissue culture technology is the adoption of practices and use of equipment to reduce the unit cost of micropropagule and plant production. In many developed countries, conventional tissue culture-based plant propagation is carried out in highly sophisticated facilities that may incorporate stainless steel surfaces, sterile airflow rooms, expensive autoclaves for sterilization of media and instruments, and equally expensive glasshouses with automated control of humidity, temperature and day-length to harden and grow plants. Many such facilities established at a high cost are high-energy users, and are run like a super-clean hospital.

The requirements to establish and operate such tissue culture facilities are expensive, and often are not available in the developing countries. For example, the cost of electricity in the developed countries is much lower, and its supply far better assured than in the developing countries. The same can be said of the supply of culture containers, media, chemicals, equipment and instruments used in micropropagation. Hence, alternatives to expensive inputs and infrastructure have been sought and developed to reduce the costs of plant micropropagation.

Low cost options should lower the cost of production without compromising the quality of the micropropagules and plants. The primary application of micropropagation has been to produce high quality planting material, which in turn leads to increased productivity in agriculture. The generated plants must be vigorous and capable of being successfully transplanted in the field, and must have high field survival. In addition, they should be genetically uniform, free from diseases and viruses, and price competitive to the plants produced through conventional methods. Reducing the cost should not result in high contamination of cultures or give plants with poor field performance. The foremost requirement of micropropagation is the aseptic culture and multiplication of plant material. Microbe-free conditions need to be maintained in culture containers, and during successive subcultures.

In many cases, mistakes in concept or practice can introduce microbes in the culture containers from an external source or the plant material itself (endophytic contamination). As a result, the microbes overgrow the cultures, and wipe them out. Microbes may grow slowly under controlled low temperature, but they proliferate very fast under uncontrolled and high temperature. Thus the adoption of wrong low-cost options may make the production process prone to disasters. Low cost techniques will succeed only if the basic conditions for tissue culture are scrupulously adhered to maintain propagule quality. Microbial contamination of cultures is known to wipe out work of months, and can turn into a nightmare.

The best low-cost option is to discard and dispose of contaminated cultures outright. Avoiding contamination in small

R&D laboratories is not a difficult task where only a low number of cultures are handled. However, commercial production involves handling of thousands of cultures each day. It is also essential to maintain such cultures in large numbers under contamination-free conditions, until they are used for either further subculture or hardening and growing-on. Plants do not have an immune system and there is a limitation on the use of antibiotics to circumvent the problem.

Moreover, many of the antibiotics, which are effective against bacteria, fungi, and phytoplasmas, are toxic to plants as well. Obviously, the use of antibiotics is not foolproof or the desired method to rid microbial contamination. Sophisticated state-of-the-art facilities are not a guarantee for prevention of contamination. The laboratories that succeed in an immaculate control of contamination do so by adherence to scrupulous techniques of basic tissue culture. Thus, it is not the sophistication but the procedures that ensure the quality of tissue cultured plants.

MICROPROPAGULAS

Low cost technology means an advanced generation technology, in which cost reduction is achieved by improving process efficiency, and better utilization of resources. Presently, both the developing and the developed countries require low cost technology to progressively reduce the cost of propagule production. In many developing countries, the potential end-users of plants derived from tissue culture have been the resource-rich farmers. They know the benefits and potential of healthy planting material. Such growers are prepared to risk investment in the high productivity potential of the planting material.

For example, hybrid seeds of many vegetables, papaya, rice, and cotton cost 15-20 times more than the price of ordinary varieties. Yet there is a wide market for them. Hence, the production of low quality plants, just because they are less costly, is not going to be a sustainable approach for the application of micropropagation in agriculture. Lowering of cost of production is possible only if the methods do not compromise the basic imperatives of tissue culture and quality of plants.

The potential of plant tissue culture in increasing agricultural production and generating rural employment is well recognized by both investors and policy makers in developing countries. However, in many developing countries, the establishment cost of facilities and unit production cost of micropropagated plants is high, and often the return on investment is not in proportion to the potential economic advantages of the technology. These problems can be addressed by standardizing agronomic practices more precisely (precision agriculture) and by achieving maximum net profits from the crops or by decreasing the unit cost of production or both.

The technology is particularly relevant to the propagation of ornamental plants. Despite high costs of production, trading of ornamental plants has thrived because they command high unit value. However, the market is limited. Over a period of time, many new tissue-culture companies in several developing countries have entered to compete in the limited market. The inevitable result has been the reduction of net margins below viable limits. Many international organizations also agree that tissue culture technology is very relevant to agriculture, provided the problem of high cost of production is satisfactorily solved.

The role of tissue culture was clearly recognized by the FAO in the paper on 'Biotechnology in agriculture, forestry and fisheries - FAO's Policy and Strategy'. The report pointed to the wide use of tissue culture techniques for multiplication of elite clones and elimination of pathogens in planting material. It also pointed to the successful tree regeneration in about 100 forest species and its value for breeding, clonal testing and rapid deployment of superior genotypes. Several R&D projects have been undertaken to improve the productivity of agricultural, horticultural and forest trees by the European Union under Co-operation in the Field of Scientific and Technical Research (COST). Under this program, coordinated and funded by the European Union, one of the primary aims has been to reduce micropropagation cost.

For example, the objective of 'COST 843' action has been the innovation of low-cost plant propagation methods that enhance sustainable and competitive agriculture and forestry in Europe. The high costs of labour of micropropagation are a major bottleneck in the EU to fully exploit in vitro culture technology. In the EU, labour currently accounts for 60-70% of the costs of the in vitro produced plants. In another program, the large-scale production and introduction of bamboo in the EU using tissue culture technology has been undertaken with the main objective of reducing the costs of micropropagation.

It has been stressed time and again that in the long-term agriculture and forestry need to be sustainable, use little or no crop-protection chemicals, have low energy inputs and yet maintain high yields, while producing high quality material. Biotechnology-assisted plant breeding is an essential step to achieve these goals. Plant tissue culture techniques have a vast potential to produce plants of superior quality, but this potential has been not been fully exploited in the developing countries. During in vitro growth, plants can also be primed for optimal performance after transfer to soil.

In most cases, tissue-cultured plants out-perform those propagated conventionally. Thus in vitro culture has a unique role in sustainable and competitive agriculture and forestry, and has been successfully applied in plant breeding, and for the rapid introduction of improved plants. Bringing new improved varieties to market can take several years if the multiplication rate is slow. For example, it may take a lily breeder 15-20 years to produce sufficient numbers of bulbs of a newly bred cultivar before it can be marketed. In vitro propagation can considerably speed up this process.

Plant tissue culture has also become an integral part of plant breeding. For example, the development of pest- and disease-resistant plants through biotechnology depends on a tissue culture based genetic transformation. The improved resistance to diseases and pests enables growers to reduce or eliminate the application of chemicals. The FAO Committee on Agriculture has perceived plant tissue culture as a main technology for the developing countries for the production of disease-free, high-quality planting material, and its commercial applications in floriculture and forestry.

It further points out that tissue culture techniques are being used particularly for large-scale plant multiplication. Micropropagation has proved especially useful in producing high quality, disease-free planting material for a wide range of crops. Tissue cultured based industry also generates much-needed rural employment, particularly for women. In forestry, the availability of tissue culture linked production systems may effectively provide sustainable alternatives to the need for harvesting wood from native forests and natural habitats.

Successful protocols now exist for the micropropagation of a large number of forest tree species, and the number of species for which successful use of somatic embryogenesis is increasing. Thus in the future, it is likely that micropropagation in the forestry sector will become commercially important. Compared to vegetative propagation through cuttings, the high multiplication rates available through micropropagation offer a much quicker capture of genetic gains obtained in forest tree breeding programs. However, the current high costs will also be one of the major impediments to the direct use of micropropagation in many programs.

The broad application of existing technologies to plantation species is important for tree improvement in the tropics. In a small number of plantation programs, micropropagation is being used as an early rapid multiplication step. However, it has been pointed out that the current high costs will be an impediment to the direct use of micropropagules as planting stock. Micropropagation clearly has a role, in the rapid multiplication of the selected clones for conventional production of cuttings. The direct use of micropropagules as planting stock in industrial plantation can dramatically broaden forestry tree farming if propagation costs are reduced.

The availability of micropropagation technologies will also be useful in genetic engineering applications, e.g., the production of plants as a source of "edible" vaccines. There are many other useful plant-derived substances which can be produced in tissue cultures, sometimes more cheaply and reliably than from natural forests and plantations. These include medicinal compounds and drugs now being sought in major prospecting operations in the tropical forests.

Micropropagation has been identified as a suitable technology in the development projects of UNESCO in Africa and the Caribbean; however, the cost of production must be reduced. Practically in all developing countries, the private industry is the most important group that requires cost-effective technology. For example, in India of the 90 commercial micropropagation units established initially, 32 were closed down. Of those engaged in commercial production, many are uneconomic mainly due to the high cost of production and the absence of quality tests. Hence, low-cost tissueculture technology will stay a high priority in agriculture, horticulture, forestry, and floriculture of many developing countries.

CULTURE MEDIA AND CONTAINERS

In vitro growth of plants is largely determined by the composition of the culture medium. The main components of most plant tissue culture media are mineral salts and sugar as carbon source and water. Other components may include organic supplements, growth regulators, a gelling agent. Although, the amounts of the various ingredients in the medium vary for different stages of culture and plant species, the basic MS (Murashige and Skoog) and LS (Linsmaier and Skoog) are the most widely used media.

During the past decades, many types of media have been developed for in vitro plant culture. Media compositions have been formulated for the specific plants and tissues. Some tissues respond much better on solid media while others on liquid media. In general, the choice of medium is dictated by the purpose and the plant species or variety to be cultured. The ratio of auxins to cytokinins in the culture medium is important since their combination determines the morphogenic response for root and shoot formation. Plant extracts such as coconut milk, banana extract, and tomato juice can be very effective in providing undefined mixture of organic nutrients and growth factors.

A variety of other media components have also been used for specific purposes. The osmolarity of the culture medium, agitation, and aeration of suspension cultures have an important influence on plant cell division. The medium can be solid, semi-solid or liquid, depending on the presence or absence of gelling agents.

Media chemicals cost less than 15% of micro-plant production. In some cases the cost may be as low as 5%. Of the medium components, the gelling agents such as agar contribute 70% of the costs. Other ingredients in the media - salts, sugar and growth regulators - have minimal influence on production cost and are reasonably cheap. However, low cost options are available to replace expensive gelling agents, sugars and reduce the cost of water. For example, the water from distillation apparatus or passing through Millipore filters can be replaced with ordinary tap water in some cases.

Gelling Agents

The growth of cultures and production of shoots or roots is strongly influenced by the physical consistency of the culture medium. Gelling agents are usually added to the culture medium to increase its viscosity as a result of which plant tissues and organs remain above the surface of the nutrient medium. Many gelling agents are used for plant culture media, e.g., agar, 'Agarose', and 'Gellan gum', and are marketed under trade names such as 'Phytagel, Gelrite', and 'Gel-Gro' (ICA Biochemicals). Agar is the most commonly used gelling agent for preparation of solid and semi-solid media. It contributes to the matrix potential, the humidity and affects the availability of water and dissolved substances in the culture containers.

Various brands and grades of agar are available commercially, which differ in the amounts of impurities, and gelling capacity. Agar brands vary widely in price, performance and composition. It is the actual use and experience, which ultimately determines the choice of agar brand in a specific system and for a plant species. It is usually unnecessary to use high purity agar for large-scale micropropagation; cheaper brands of agar have been successfully used for industrial scale micropropagation.

The lowest concentration of agar, which can be used, depends on its purity and brand. Agar is usually used at 0.6-0.8% (w/v). It is advisable to prepare sample media in small quantities using various concentrations of agar, e.g., 0.7, 0.75, 0.8, 0.85 and 0.9%. The appropriate concentration should then be used for large-scale production purposes. In addition to cost saving, there are a number of other advantages in using low concentrations of gelling agents. A semi-solid medium ensures adequate contact between the plant tissue and the medium. It is beneficial to growth as it allows better diffusion of medium constituents, and is easily removed from plantlets before their transfer to in vivo conditions. For these reasons, a semi-solid medium is often preferred over solid medium.

Alternatives to Agar

Cheaper alternatives to agar include various types of starches and plant gums. The National Research Development Corporation, India (NRDC) has listed low cost agar alternatives, which are worth evaluating for routine use in commercial micropropagation. Gelrite can be replaced with starch-Gelrite mixture. The use of liquid media eliminates the need of agar. Other options include white flour, laundry starch, semolina, potato starch, rice powder and sago. For micropropagation of ginger and turmeric, the combination of certain gelling agents gave growth as good as on agar-based media.

The use of laundry starch, potato starch and semolina in a ratio of 2:1:1 reduced the cost of gelling agent by 70-82%. However, the addition of such gelling agents to the medium also has some disadvantages. Some gelling agents contain inhibitory substances that hinder morphogenesis, and reduce the growth rate of cultures. Moreover, toxic exudates from the cultured explants may take a longer time to diffuse. Media solidified with gelling agents increase the time to clean the culture containers. The low cost options to agar, agarose, and gellan gum are listed below.

Corn-starch (CS) as a gelling agent has been used along with low concentration of 'Gelrite' (0.5 g 'Gelrite' + 50.0 g CS /l) for the propagation of fruit trees, such as apple, pear and raspberry, banana, and sugarcane, ginger and turmeric. The shoot proliferation was better on corn starch-medium than on agar. The cost of CS was $1.8/kg compared with $200/kg of agar. However, it became difficult to detect the contamination because the CS medium turned grayish-white. Addition of 8.0% tapioca starch to the MS medium was found to be a good substitute for 'Bacto-agar' for potato shoot-culture.

Barley starch (60.0 g/l) has also been used for culturing potato-tuber discs, and for anther culture of barley. Sago (obtained from the stem pith of Metroxylon) at 13% concentration was substituted for agar in MS medium for the multiplication of chrysanthemum. The number of shoots and leaves, and root length were significantly higher on sago than on agar. The cost of sago is $0.5/kg. 'Isubgol', a colloidal mucilaginous husk (chiefly composed of pentosans) derived from the seeds of Plantago ovata), has a good gelling activity, and has reasonable clarity in gelled form. 'Isubgol' at 3% in MS medium has been used for the propagation of chrysanthemum. The cost of 'Isubgol' is about $4/kg.

Suspension cultures without gelling agents are commonly used for culturing callus, cell clusters, buds and somatic embryos. Suspension systems allow greater contact between the explant and the medium. Moreover, agitation of such media reduces the diffusion gradient in the nutrient supply. The toxic metabolites exuding from the tissues are also dispersed effectively. Sterilized, non-chlorine bleached, rolled, pure cotton fiber has been successfully used for culturing callus and proliferating shoots of Taxus, Agrotis and Artemisia.

Recently, the use of cotton fiber support to cultures in liquid media has been reported in the commercial propagation of orchids, banana, chrysanthemum, and potato in Pakistan and Bangladesh. Cotton fiber and MS liquid medium produced rapid growth of banana shoots, which could be sub-cultured after two instead of six weeks. Protocorm initiation and shoot development of orchids was much faster on cotton than on agar-based medium. The cost of cotton fiber is about $2/kg, and of agar $100-200/kg depending on the manufacturer). Other alternative culture supports include foamplastic, filter paper bridges, glass beads, 'Viscose' sponge, glass wool and rock wool in liquid media.

Strips (2.5x15 cm) of glass wool, nylon and filter paper have been used as supporting bridges for the propagation of chrysanthemum on MS medium. The shoot- and root-growth was almost identical on these matrices. Polystyrene foam blocks (2x2x1 cm) have been used as supporting matrix for the propagation of chrysanthemum. The growth response was similar to that on glass wool, except that explants on foam produced significantly fewer roots but longer shoots than on glass wool.

Glass bead based liquid-media were successfully used for culturing ginger and turmeric, and on per plant basis reduced the medium cost by 94% With glass beads, the amount of medium required is only 15-18 ml per culture container. By using 20ml medium per culture container, one liter media will give 50 culture containers, a substantial saving in medium cost.

Ginger and turmeric plants multiplied on glass bead liquid-medium performed as good as on agar-based medium. A similar type of response was observed for vanilla. Ficus cv.'mini lucii' showed higher multiplication rate although with a slight vitrification. Saintpaulia, Syngonium, Philodendron and Spathiphyllum also showed higher multiplication rates and better growth on glass bead liquid-medium. An additional advantage of liquid media with glass beads is that the medium can be replaced without moving plants.

Glass beads have been used for the propagation of raspberry and white clover resulting in 60% saving in media cost. The glass beads allowed easy removal of plantlets from the medium. Glass beads have been also used for the maintenance of callus and shoot organogenesis. The beads can be reused after washing with acid. The liquid-media have some disadvantages. Delicate tissues get damaged during agitation. In some species, shoots submerged in liquid media become hyperhydric (water soaked), and unsuitable for micropropagation.

Carbon Sources in the Micropropagation

Sucrose is the most commonly used carbon source in the micropropagation of plants. Sucrose adds significantly to the media cost. Household sugar and other sugar sources can be used to reduce the cost of the medium. Sugar sold in grocery stores is sufficiently pure for micropropagation. For culturing ginger and turmeric, all other carbohydrates except sugarcane juice, were suitable alternatives to laboratory grade sucrose. Use of common sugar reduces the cost of the medium between 78 to 87%. The cost of the local sugar was US$ 0.55/kg against the $40.0/kg for the imported sucrose.

Several other ingredients can also be replaced by low cost options Commercial grade chemicals of lower purity than the analytical grades are quite suitable for commercial micropropagation unless deleterious effects are observed. A high degree of purity is justified only in the case of basic studies in tissue culture. In general commercial micropropagation, the quality will hardly be affected ordinarily by purity of these chemicals. Of all the chemicals, growth regulators (hormones) are the most expensive. However, they are needed in very small amounts in the medium, thus have a little effect on the medium cost.

Sugar cane molasses can provide many of the nutrients,namely, sugar, vitamins and inorganic metal ions required for sugarcane callus induction and shoot formation. At Dhaka University, low cost media have been developed for the multiplication of orchids based on macro-salts of any of the Vacin and Went, Phytamax and Knudson's C media, supplemented with 10-15 % (w/v) banana extract and 10-15 % (v/v) coconut water. Use of above media eliminated the need of micro-salts and vitamins for the micropropagation of different orchid species. Several different orchids, namely, Vanda, Dendrobium, Aerides, Acampe and Spathoglottis can be multiplied on a medium containing only peptone, inositol, banana extract and coconut water. In many tropical orchid species, similar response was observed on media containing coconut water (15%) and banana pulp (100 g/l), which effectively lowerd the cost.

Importavce of Water in Tissue Culture Media

Water is the main component of all plant tissue culture media. Usually in tissue culture research, distilled or doubled distilled and de-ionized water is used. Distilled water produced through electrical distillation is expensive. In some cases, alternative water sources can be used to lower the cost of the medium. If tap water is free from heavy metals and contaminants, it can be substituted for distilled water. Tap water has been used for in vitro propagation of banana and ginger, Zingiber officinale. Table bottled water from the supermarket can also be used a low cost alternative.

However, its mineral composition should be taken into account as it may affect pH and nutrient uptake. In rural areas, rainwater can be collected in clean glass jars and used for tissue culture. In Bangladesh, the change over of water distillation from electrical to gas operated unit reduced the cost from US$260 to $5/month for producing 50-60 liter water per day.

Autotrophic Micropropagation

Plants with functional chloroplasts can grow in vitro on media without sugar, provided the micropropagation environment is modified to enable photosynthesis. The growth of plants on sugar-free medium, but with the carbon dioxide enriched environment was similar to sugar-containing media. Plants such as carnation, chrysanthemum, Cymbidium, Primula, potato and strawberry have been successfully grown using the above technique, termed as PTCS-'Photoautotrophic Tissue Culture System'. In the autotrophic system, plants are grown in large containers where the air content (oxygen, carbon dioxide, relative humidity, etc) and the composition of the culture medium is easily controlled. This system has several advantages that include reduction in production cost because of large culture vessels, simple culture media formulations, and lower incidence of culture contamination. Plants grown in such strongly aerated vessels require little or no hardening.

TYPES OF CULTURE CONTAINERS

A wide variety of culture containers are available on the market.

The plant performance and cost of the containers should be used as the prime criterion in choosing the appropriate type. Depending upon the scale of production, different types of containers are deployed for culture initiation, maintenance of mother cultures and sub-culture for multiplication. For shipment of plants, disposable containers should be bought as and when required. Irrespective of the type, the containers used for maintaining in vitro plants should be transparent to facilitate illumination and easy inspection.

Glass test tubes have been universally used for culturing plant tissues as their narrow openings keep out contamination. Usually, only 1 to 2 explants are cultured in each test tube. As explant growth and multiplication progress, the cultures are transferred to larger containers. Test tubes should not be used for large-scale multiplication of plant material. Conical flasks, 150-250 ml capacity, are mostly used for the micropropagation of plants in liquid media, and are kept on shakers for agitation of the medium and suspension cultures. However, conical flasks are more expensive than glass bottles, and their narrow mouths make manipulations of cultures difficult.

Glass and plastic Petri dishes are also used to culture explants, which are then transferred to larger vessels for multiplication and elongation. Glass Petri dishes have to be sterilized before pouring medium that has to be sterilized separately. Glass Petri dishes are more expensive than the test tubes and conical flasks. Pre-sterilized disposable-plastic Petri dishes are much cheaper. In both cases, medium is poured under the laminar flow hood.

Glass bottles and baby-food jars with polypropylene caps are the most widely used containers and the most economic and low cost option. The wide mouth makes culture manipulation easy and approximately 15-20 explants can be inoculated in each bottle. Such containers are widely available and cost ca. US$0.09. Such containers can be washed, sterilized and reused repeatedly. In Cuba, glass containers are washed with a circular nylon brush mounted on an electric rotor, and then dried in the sun by inverting them on wooden trays. The containers are then double wrapped in newspapers and sterilized in an autoclave.

The medium is sterilized separately in flasks, and then dispensed into the containers under the laminar flow cabinets. Transparent plastic containers, such as 'Magenta'TM vessels, that withstand autoclaving and washing, are extensively used for micropropagation in many developed countries, but at US$1.5 to 2.0 are very expensive. Also, the repeated autoclaving of the plastic containers renders them cloudy, thus reducing the passage of light. Lately, containers are being made of polypropylene and polycarbonate and more recently polystyrene that can be autoclaved.

Disposable, non-autoclavable food containers and sandwich boxes made from polystyrene have been also used for plant micropropagation. Use of disposable containers eliminates the cost of washing. PVC pots and jars, manufactured for the food-industry, have been used for in vitro plant culture. The high temperature during their manufacture imparts relatively high degree of sterility, so that they can be used directly without sterilization. Hence, only the medium needs to be sterilized, thus eliminating the costs of container sterilization. Non-sterile disposable plastic type containers and bags can be bulk sterilized with gamma rays at industrial irradiation facilities. Such gamma-irradiated containers are being used for large-scale micropropagation. Being lightweight material, these can be shipped to distant places at a very low cost.

A new type of culture vessel called 'StarPac' TM has been manufactured. It is a disposable bag, and is semi-permeable to gas but requires different methods of handling in contrast to test tubes and other containers. Disposable plastic type containers called 'Watson Modules'TM, have been used for production of micro- and mini-tubers of potato. The modules allow in vitro multiplication, plant hardening and soil growing in the same container, and permit batch handling of cultures during all stages of plant growth. Disposable containers have been also made from fluorocarbon plastic films, which can be autoclaved, and are transparent and impervious to water, but fairly permeable to gases.

Vessel Closures

The vessel closures influence growth of the in vitro plants. The culture containers have to be closed to keep out microorganisms but they should not be sealed so tightly that gaseous exchange is prevented. The different types of closures include non-absorbent cotton plugs, Polyurethane foam plugs, specially made plastic plugs, aluminum foil, stainless steel caps, Polypropylene caps with bends, PVC film, Polythene film and silicon rubber. Culture containers such as tubes and flasks are closed with non-absorbent cotton wool, sometimes wrapped in muslin gauze. But under large-scale production, preparing cotton plugs becomes cumbersome and time consuming. Hence, these have been replaced with autoclavable screw caps made of stainless steel or polypropylene.

Test tubes covered with translucent caps, which facilitate good gaseous exchange, have been used for culturing carnation. Polypropylene film, which is heat resistant, has been successfully used as closures. It is transparent, non-greasy, and can be used to seal large containers. Sandwich wrapping films, such as 'Cling film' has also been used either on its own or as double wrap on top of the lids to prevent contamination after subculture. This is particularly suitable for mother culture, which may have to be maintained for 3 to 6 months or longer.

Gaseous exchange is important for the availability of oxygen and carbon dioxide, water vapour and elimination of ethylene gas build-up that is detrimental to plant growth. The shape of the container influences the growth rate of cultures by modifying gaseous diffusion. The best size and shape of the culture container varies with the plant species, and hence should be decided after prior experimentation. It is important to ascertain the price, costs for washing, and the number of times the culture container can be reused. The selection of the containers should be done after making sure that they would not affect growth in vitro. Culture containers can be vented by modifying lids. to facilitate optimal growth. Depending on the plant species, microporous membranes of different sizes are available.

Plastic bags (approximately 10x15cm) have been used for large-scale micropropagation and are very cost effective. Disposable plastic bags eliminate the cost of washing and of lids. Plastic bags are sterile due to high temperature during manufacture. After pouring presterilized medium under the laminar flow, the top 2 to 5 cm of the bags is folded and several bags are held together with either large paper clips or plastic cloth-hanging pegs. After transfer of the explants and cuttings, the bags are closed with a heat-sealing machine or by knotting if the bags are 18-20 cm long. Plastic bags being lightweight can also be hung by thread and do not need elaborate shelving. In Bagladesh and India, juice, jam, and jelly bottles have been used. The orchid producers in Thailand, Singapore, Indonesia, Malaysia, and India have been using old whisky bottles for orchid culture for many years.

DISEASE DETECTION AND ELIMINATION

The health status of the donor mother plant and of the plants multiplied from it are among the most critical factors, which determine the success of a tissue culture operation. Hence, indexing of the mother plants for freedom from viral, bacterial, and fungal diseases is a normal procedure before undertaking propagation in large-scale plant propagation through tissue culture.

Fungal and Bacterial Diseases

While most of the fungal and bacterial diseases are eliminated during surface sterilization and culture, viruses and viroids survive through successive multiplication if the mother plant is infected. Even cultures from seeds may carry viroids and bacterial endophytes, in certain cases also viruses. Bacteria may be controlled under in vitro conditions to some extent using certain bacteriostatic agents. Nevertheless, such chemical treatments with e.g. 'plant protection mixture' (ppm) or commercial antibiotics even when applied on long term do not fully eliminate microorganisms.

Viruses are obligate parasites, and for their multiplication, they use the metabolic machinery of the cells, and inhibit plant growth. Although, many viruses do not produce disease symptoms, they adversely affect plant metabolism and increase progressively with repeated vegetative propagation. Most viruses infect only a limited number of species, but a few have a wide host range. Many plant viruses are transmitted through vectors such as insects, nematodes, bacteria, and fungi. Propagation from vegetative parts, grafting and sap can transmit viruses from one plant to another. The damage from viruses depends on their concentration in the tissues and spread within the plants.

The greater the concentration of the virus, the greater the debilitating effect observed on plant growth. Many varieties of vegetatively propagated crops decline in performance with viral accumulation and must be discarded. Even though the viruses do not produce visible symptoms, the plants have reduced yield and poor quality. The removal of specific viruses by meristem-tip culture has been reported to lead to a dramatic yield increase and rejuvenation of plant varieties that are vegetatively propagated. This is a unique contribution of meristem-culture not achievable by any other techniques.

Plants not originating from pathogen-tested material must be screened for the presence of viruses. Procedures to detect the presence of viruses include visual examination for viral symptoms, infection tests on indicator plants, serological tests, electron microscopy, and direct detection of RNA using molecular techniques. Internationally approved and recognized detection systems include serological and molecular laboratory assays, and indicator hosts in the greenhouse.

Prior to the advent of antigen-antibody reaction detection systems, the earlier method of detecting virus relied upon indicator plants. For example, in potato, viruses A, X, and Y can be detected by rubbing the juice of the infected plant on the leaves of indicator plants such as Chenopodium spp., and Nicotiana spp. The development of necrotic spots on the leaves is a typical symptom indicating the presence of the viruses. For example in sweet potato, its wild relative, Brazilian Morning Glory (Ipomaea setosa) is still used as the indicator host.

Similarly, in many fruit trees, the appropriate rootstocks are used for indexing of various virus strains. The detection of plant diseases has changed radically in the last 20 years. A number of indicator plant based tests have been replaced with specific Enzyme Linked Immuno Sorbent Assays (ELISA). Nevertheless, indexing with indicator plants is a low cost and highly effective method to detect many viruses. ELISA has been the most effective method for virus and pathogen detection in plants.

Each virus has a unique protein coat. Hence they can be detected by using antibodies, which are usually linked to enzymes for reporting the affinity binding via colour signal from the used enzyme substrate. The double antibody sandwich (DAS) method in a 96-well microtitre plate is still widely used. Various alternative formats, substrates, and antibody binding modifications have been developed over the years to increase specific sensitivity. The most common method nowadays uses "biotinylated" virus-specific antibodies to improve their affinity to the respective virus surface thus improving the signal.

The use of monoclonal antibodies has been so far restricted to certain potato and barley viruses, because of their high cost and inability to detect the range of strains in several diseases.

Large scale testing based on screen-spot samples is also a routine procedure. Such dotimmuno-binding assays are versatile and cheaper alternatives to microtitre plate tests. Plant sap is directly spotted onto a pre-treated microporous membrane (e.g. nitrocellulose or polyvinylidene difluoride). The test kits can be used directly in the field. Sample membranes can be dried and shipped to another laboratory for processing. However, such systems are not useful for viruses present in low concentrations in the plant sap.

ELISA has several limitations. First of all, a test is specific only to a given strain of the pathogen. For a large group of ubiquitous viruses, which do not produce disease symptoms or fatal consequences, ELISA kits have not been developed. Thus, the strain specificity has severe limitations for comprehensive quality control. Further, ELISA is less sensitive than PCR, and may fail to detect low amounts the virus in tissue-cultured plants. Whereas a positive ELISA is a good indicator of the existence of microbes (but false positives may also occur), a negative test does not ensure their absence. The use of Immuno-Tissue-Printing allows the localization of viruses in tissues and therefore the improvement of elimination strategies in vitro by visualizing the result of virus removal.

Polymerase Chain Reaction (PCR)

The PCR is more sensitive than ELISA, and can detect pathogens in extremely low amounts. PCR detects pathogens from their genetic material, i.e. DNA or RNA. This technique uses a specific enzyme and the respective genomic start code for a relevant section of the pathogen-DNA to reproduce millions of copies from it. Therefore, even minute quantities of the pathogen DNA can be detected. Since most plant viruses are made of only RNA, first a DNA copy of the RNA is made, whereas DNA from bacteria and phytoplasmas may be amplified directly. Because the genetic material of the pathogen is mixed with that of the plant, it is necessary to target only the genetic material to be copied. This is done by adding small pieces of DNA called primers which stick only to borders of the region of genetic material that need to be copied.

The primers then direct the enzyme where to start and end amplification. During PCR, the DNA is subjected to a series of hot, cold, and warm cycles in a thermocycler. Hot cycles split DNA into single strands. Cold cycles allow the primers to attach to the DNA. During warm cycles, the enzyme makes a copy of each piece of primed DNA. If a phytoplasma is present, PCR produces a large quantity of a specific piece of DNA. By using gel electrophoresis, DNA pieces are separated in size, and the specific piece of DNA is identified.

If the critical concentration of a DNA particle is sufficiently low and the sample tests negative, the plant material can be regarded as practically free of the target microbe. Once a test is standardized with an appropriate primer, and by a method, which excludes inhibitors, a negative PCR test is a more reliable indicator of practical freeness. The use of PCR drastically reduces the number of samples, which test negative although they may carry the virus. Once proper probes are developed, the nylon membrane spot test is a practical, rapid and accurate method for diagnosis. PCR has the advantage that by determining "degenerate primers", a fairly large group of microbes such as viruses, bacteria and fungi can be detected.

Nucleic acid Hybridization Test

The presence of viruses can be detected from isolation and characterization of double stranded RNA (ds-RNA) produced in most plants during viral replication. Although nonspecific presence of ds-RNA in a sample extract strongly suggests viral infection, for further identification of the virus, the test has to be verified by serological and other methods. Viroids are much smaller than viruses, and are made of a circular piece of RNA and do not have a coat protein. Viroids such as 'Potato Spindle Tuber Viroid' can be detected by nucleic acid hybridization.

The RNA of the virus "hybridizes" with probe RNA. When the two single strands come together, they form a double strand. To detect the viroid, RNA extracted from plant tissue is attached to a microporous membrane. This bound RNA is then detected by a "probe", which is an RNA particle labeled with either 32P or a chemical called DIG (digoxigenin). If a viroid is present, the probe hybridizes with it; if not, the probe is lost when the membrane is washed. DIG bioluminescence or radioactivity if present on the membrane can be detected on an exposed film. Samples from known healthy and known infected plants are used as controls. Large scale testing on nylon screen-based spot samples is also a routine procedure.

Viruses Elimination

The elimination of viruses can be achieved by a combination of apical meristem culture and thermotherapy. Meristem culture is the most commonly used method to free plants from virus diseases. This should not be confused with shoot tip culture as explained before. Virus-free plants have been obtained by thermo-therapy and meristem culture in several species. The plant meristem is a zone of cells with intense divisions, situated in the growing tip of stems and roots. The virus travels through the plant vascular system, which is absent in the meristem. Moreover the cell-to-cell movement of the viruses through plasmodesmata cannot keep up with the growth and elongation of the apical-tip.

The high metabolic activity of meristematic cells, usually accompanied by an elevated endogenous auxin content in shoot apices may also inhibit virus replication. Thus, the meristem is highly protected from infection. Based on this finding, meristem culture has been extensively used to eliminate viruses, bacteria and fungi from plants. The culture of meristems or alternatively small shoot tips, in combination with enhanced cell division in vitro and/or thermal pre-treatment allows the elimination of viruses in plants propagated from vegetative parts. If explants are too big, they are likely to contain virus particles in the associated vascular tissue.

For thermotherapy, the plants are first are grown at high temperature for 4-6 weeks. Under tropical or subtropical conditions, this can be accomplished simply by installing a small compartment of a glasshouse equipped with a roof vent on one end and an exhaust fan on the opposite end, both temperature-regulated. This approach removes the excess heat and provides a constant high temperature daytime treatment. In climates with temperate conditions, the same effect can be achieved by placing fluorescent lights including ballasts and/or heat producing incandescent lamps in the necessary minimum distance to avoid damage over the plants to be treated in a dark box just large enough to include the plants. Such a system has been used for virus elimination in sweet potato.

After thermotherapy, 0.2 -0.4 mm explants are preferentially cultured singly in test tubes. If the explant is too big, it likely has a vascular system that may contain microbial contaminants including viruses. The plants thus obtained are multiplied, and re-indexed. A better strategy is to culture 2-5mm long explants for 4-5 weeks, maintain the in vitro grown plant at high temperature for 4-5 weeks, and excise 0.2-0.4mm or even longer explants to initiate subcultures. This procedure avoids in vivo contamination problems and gives high survival and multiplication rate.

In potato, in vitro cultures were established from several millimeter long shoot tips and axillary buds after in vitro thermotherapy to free potato from all relevant virus diseases. By using this method, it was possible to eradicate viruses A, Y and X from potato varieties in one single step (Ahloowalia, unpublished). A method called multiple lateral shoot technique of in vitro elimination of three common viruses, X, Y and S of potato, has been also reported. In this method, a stem with at least five nodes is treated with ribavirin in liquid medium, and cultured for 5 days at room temperature.

The stems are subjected to thermotherapy at 32-350 C for 25 days, apical buds are taken from the lateral shoots cultured on solid medium, and checked for virus infection with ELISA. The survival of explants in relation to the applied temperature stress and their size is always inversely proportional to the pathogen elimination success. It is known from molecular and traditional microbiological surveys that effectively all in vitro plants contain certain microorganisms, mostly bacteria but also mycoplasmas, viroids, and fungi, many of which cannot be cultured without the host. Such microorganisms may not show symptoms, and in some cases may even have a positive effect on the growth performance of the host plant.

Shoot-tip culture is used for the multiplication of plants that are already freed from known diseases. It should be emphasized that it does not free the plant from viruses. In fact, it favors propagation of viruses and increases virus concentration in the daughter plants. In many ornamentals, variegation is due to the presence of certain viruses or mycoplasms. The removal of such viruses is therefore not desirable. In such ornamental plants, multiplication from shoot-tip and axillary bud culture is the better choice, because meristem culture may remove the virus, and the ornamental variegation is lost.

In certain varieties with variegations as in Petunia and Pelargonium meristem culture may dissolve the existing chimera, thus producing material without the desired character. Viruses may also produce certain phenotypes that are characteristic of a plant variety. For example, in sugarcane cv. 'Co 740', the yellow irregular leaf streaks were initially considered as a diagnostic character till they were proven to be disease symptoms.

The removal of the virus by meristem-tip culture led to the disappearance of the yellow streaks, which reappeared on re-infection. Erroneously, the yellow streaks are still recognized as a varietal diagnostic trait of 'Co 740'. Recently, endogenous para-retroviruses have been reported that integrate in the genome of the plant material. Under in vitro stress, sequences of such viruses have been detected in the micropropagated banana plants.

Despite the fact that many viruses may not show visible symptoms, their presence can reduce the yield and quality of crops. Yield increases of up to 300% were reported following replacement of virus-infected stock with specific pathogen-free plants. In China, yield losses from virus diseases of sweet potato (Feathery Mottle Virus, SPFMV, Latent Virus, SPLV and Caulimo-Like Virus, SPCLV) used to be as much as 20%. After the establishment of a virus elimination program, within a few years almost 10% area of 6.6 million ha under sweet potato is now planted with virus-free planting material. There are no chemicals to cure virus-infected plants in the field.

However, some less virulent viruses can protect plants from infection of their more virulent strains and create cross protection. The phenomenon of cross protection need not prevent the use of meristem culture. Under field conditions a crop does not stay absolutely virus free even though derived from tissue culture material. Even under tissue culture, total freeness may not be achieved. Many symbiotic organisms contribute to the vigor and ability of the tissue-cultured plants to perform well in the field. In the course of tissue culture, many plants are made free from such beneficial microbes, e.g. symbiotic nitrogen fixing endophytic bacteria, and mycorrhizae. There is no practical way to retain the beneficial microorganisms during tissue culture. The deliberate re-infection of propagules with selected strains can be a valid way of retrieving the benefits of such microbes.

QUALITY ASSURANCE

Quality assurance becomes even more important when the plant material is for export. High quality of the super-elite and elite micro- and mini-tubers of potato has improved the production of quality seed potatoes in Ireland and Scotland where conventional multiplication had been inadequate to generate high quality disease-free planting material. Cost reduction based on low cost options is necessary for commercial micropropagation. While cost reduction is necessary for commercial micropropagation, the low cost option should not compromise quality. The sturdiness of plants is an important attribute of quality, which determines their ability to adapt to field conditions.

The cost-effective systems should produce semi-hardened plants in the culture itself. Such plants produce a profuse and viable root system and adapt in less time to field conditions. Quality standards require the establishment of suitable tests to maintain quality control and the assurance of micropropagated material irrespective of the system used. No industry can prosper without a quality policy based on carefully identified, relevant and recognized quality specifications and standard methods to test and verify them. Hence, the plants coming out of any facility should be subjected to quality testing without discrimination or prejudice.

The commercial micropropagation industry, which is still in its infancy and early stages of evolution in many developing countries, lacks this vitally important component. Under the seed law, tissue culture plants are defined as "seed". The specifications stipulated for seeds of cultivated crop plants do not apply to tissue culture plants for quality, except for the basic requirement of distinctiveness of the variety, genetic uniformity, and stability. This lack of fundamental specification creates conflicting opinions on the quality of micropropagated plants. Such a situation is undesirable for users, producers and regulating agencies and a hindrance to the maturity and growth of commercial micropropagation as a vibrant industry.

Several commercial micropropagation laboratories already follow standard practices such as labeling of cultures, use of indexed stock mother plants, and preventing secondary infection from pathogens during the hardening phase. The quality considerations need to take into account several parameters as a package. A comprehensive approach is needed, which will help micropropagators to monitor production quality, and formulate a quality- assurance system based on such tests and specifications, that are open to third party inspection. Such an approach will also be useful to formulate well-defined quality specifications by regulatory authorities.

The explant source (e.g. meristem tip, axillary buds), freedom of the donor plant from viruses, disease causing fungi, bacteria, viroids, phytoplasmas, vigour and conformity of the variety, and elimination of somaclonal variants are critical for maintaining plant quality. Many cultures are initiated from meristem-tip explants. Such cultures can be multiplied by enhancing axillary branching, which preserve genetic stability even on indefinite subculturing. Since this method reduces the chance of differentiation of plants from cells other than the meristematic tissue, the expected genetic variation is low.

However, the shoot-tip and bigger explant cultures have differentiated cells, in addition to the undifferentiated meristematic cells, and may result in somaclonal variants. Thus, shoot tip culture has a significant probability of producing genetic variants. However, in banana, the frequency of such variants markedly differs with the variety. Shoot-tip explants infected with endophytic microflora and viruses transmit them to daughter plants. Endophytic microflora is not desirable, unless it has symbiotic functions. Hence, meristem culture is the preferred method to initiate cultures for producing high quality plants.

Multiplication should be based on enhanced axillary bud proliferation, i.e. the rapid formation of tiny microscopic axillary buds in the axils of the leaves of the tiny plants that sprout and develop into daughter plants. The introduction of callus phase usually stimulated with growth regulators such as 2,4-D will introduce unwanted somaclonal variation and should be avoided during multiplication. For the same reason, the use of relatively high concentrations of growth regulators such as BAP, IAA, & BA should be avoided.

Quality assurance begins with culture initiation, explant source, and judicious use of growth regulators during multiplication. Cultures initiated from meristem tip and multiplied through enhanced axillary branching avoid callus formation and give uniform clones. On the other hand shoot-tip cultures are considered inferior to meristem-tip culture since they can be the source of virus infection. In case of photoperiod-sensitive plants, light duration can also have a serious effect on the final performance of the micropropagated plants, particularly those meant for flower production. Hence, a neutral photoperiod of 12/12hr light/dark is preferred during multiplication.

Disease Free Plants

It is necessary to initiate cultures from the meristem-tip and ensure further multiplication free from viruses. For example, the badna virus in banana integrates into the host genome during tissue culture and can become a source of infection in micropropagated plants although parents are free from symptoms. If the parent plant is virus infected, the cultures from its shoot-tip will likely be infected. Even the meristem tip will carry over viruses unless thermo-therapy has been carried out and the cultures have been tested virus free. Removal of viruses will be positively undesirable in exceptional cases of ornamentals, where the characteristics are associated with the presence of viruses, as in the case of the yellow vein geranium.

The quality tests should be rapid, and aim at the identification of a variety or clones, and ignore point and bud-mutations, which occur in the normal population as a natural phenomenon. A variety is an assemblage of closely similar genotypes and clones, and not a group of "absolutely uniform" genotypes. Thus, the methods should be able to make a distinction between two varieties, and not produce confusion with regard to intra-varietal variability.

The tests for quality include endophytic bacterial count. Broad range identification methods for the presence and quantification of endophytic bacteria are required and should be the first tests in quality control. This will prevent production of plants contaminated with undesired endophytic bacteria, and may indirectly be a useful indicator of the absence or of low concentration of viruses. Conventional bacteriological methods are enough to standardize such a test. PCR based methods may provide identification of a wide range of bacteria. For the identification of viruses, light microscopy for inclusion bodies has limited sensitivity. Immunology based methods have been commonly used for detecting specific viruses e.g. in potato. However, PCR methods are being increasingly developed and finding increasing use.

Culture Identity

Micropropagated varieties in culture and at the hardening stage are difficult to distinguish visually from one another. Variety identification by proper labeling at all stages is essential to ensure varietal identity. When several varieties of the same plant species are used, accidental mislabeling or mixing may occur. Therefore, monitoring is essential to detect admixtures at all stages. Clones of the mother variety should be maintained by conventional vegetative propagation to fall back upon whenever necessary.

Varietal conformity based on morphological characters has been the basis for identification in conventional propagation and the same applies to the plants derived from tissue culture. However, considerable caution is advised in selecting morphological characters for varietal conformity. For example, in sugarcane, variation in morphological characters such as cane colour, shape of the bud, size of the ligule can vary in the different parts of the same cane. Where morphological characters of mother plants of diagnostic importance are due to interaction with microbes, meristem tip culture invariably gives rise to a new phenotype due to removal of such microbes.

In such cases, the new virus free phenotype represents the original expression of the new virus free phenotype. Hence, criteria other than morphology are necessary for varietal conformity. Morphological characters are not an absolute proof of genetic conformity or genetic stability in tissue culture. Many times, physiological changes may lead to the appearance of types among micropropagated plants that differ from the parental clone. In some cases, such variants with minor phenotypic changes can be used for field production, but not for propagation. Low frequency mutations and bud mutations occur in conventional populations propagated from vegetative parts. Genetic changes not associated with morphological traits can be detected by DNA based molecular methods. Somaclonal variants are undesirable in cloning, but can be used for breeding.

DNA Based Molecular Markers

DNA based molecular markers are not affected by the environment and are more reliable than morphological characters for identification of varieties but are not a low cost option for small facilities. Hybridization and RAPD (Randomly Amplified Polymorphic DNA) based methods can discriminate between varieties and clones. However, the differences between the donor plant and those regenerated through tissue culture cannot be detected with such methods. Only in a few cases, it is possible to discriminate the mother plants from the variant regenerated plants. Molecular methods appear to be reliable for discrimination at the variety level.

Variety identification based on DNA fingerprinting, such as RAPD, microsatellites, STR's (Sequence Tagged Repeats), AFLP (Amplified Fragment Length Polymorphism) and multilocus probes, is potentially useful. RAPD can be standardized within a short time without elaborate knowledge of DNA sequence. However, limitations in the reproducibility of this method (between two sets of experiment, two machines, two workers and two laboratories) are well recognized. This limitation can be overcome by including a sample of reference DNA. Epidermal patterns and structures and stomata, if stable in tissue culture, are additional tools in variety diagnosis. If their validity is confirmed, they may be simpler than RAPD and other molecular methods.

PRIMING TISSUE CULTURED PROPAGULES

The adaptation of plants to the environment - temperature, osmotic stress, and pathogens, is triggered by modification of expression of specific genes and metabolic pathways. Certain chemicals, environmental factors, and microorganisms pre-sensitise the cellular metabolism of plants. Upon exposure to stress, the pre-sensitised or primed plants adapt better and faster than the non-primed plants. The production of high quality and vigorous plants through in vitro culture requires the enhancement of posttransplanting ability for water management, efficiency of photosynthesis and resistance to diseases.

Priming of in vitro propagules, based on manipulation of the growing environment (chemical, physical and biological) prior to and upon transplanting, is an integral part of tissue culture propagation. The ability of the propagules to withstand transplanting stress very often determines the success or failure of tissue culture operations. Tissue-cultured propagules are produced in controlled environments under pre-set temperature and low-light intensity, and in small vessels with poor air exchange. This leads to high humidity and accumulation of ethylene.

Also, the nutrients in the growing media, sucrose, mineral salts, vitamins and growth regulators, are far in excess of those in the soil. Such an environment causes developmental distortions, and represses or modulates several metabolic pathways. As a result, the plantlets have small juvenile leaves with reduced photosynthetic capacity, and malfunctioning stomata. The roots of such plants lack hair or have few, and the leaves have poor cuticle development and low wax deposit.

The ability to manage water, photosynthesis and response to stress during and after in vitro culture determines the final performance of the propagules. The formation of a functional root system is an essential step to manage water under the post vitro conditions. The in vitro roots remain functional and continue to grow during ex-vitro acclimatisation. Easy to root species, (e.g. potato) usually regenerate roots upon subculture on hormone-free medium. However, root architecture and biomass partitioning between roots and shoots can be easily manipulated by simple adjustment to the growing conditions that affect the post-transplanting performance of the propagules. For example, increasing day/night temperature from 20/15 to 33/250C either slightly stimulates or has no effect on the shoot growth in clones of potato, but dramatically enhances root growth and root branching.

Several chemicals also influence the development of the root system. Abscisic acid is known to affect initiation of new lateral roots and overall changes to the root architecture. Jasmonic acid (JA) has a similar effect. Low JA concentrations (0.5-1.0μM) stimulate both root and shoot growth. Jasmonates induce stress resistance in plants, and may be useful for priming of in vitro plants during the last subculture. In potato, nodal explants from JA primed stock plants tuberize earlier and more uniformly, and produce more microtubers than the non-primed controls.

Salicylic acid (SA) is a major signalling compound and induces abiotic and biotic stress resistance in plants. It also induces adventitious roots in cuttings and stomata closure, and inhibits ethylene biosynthesis in detached leaves; hence it may also be useful for priming tissue-cultured propagules. Many woody and herbaceous perennials require auxins for rooting. Auxins are involved in root initiation and adventitious root formation. However, excessive auxins after root induction may inhibit root regeneration and elongation, and impair formation of a good root system. Therefore, a careful selection of the auxin type and concentration is recommended.

IAA and IBA are usually used for easier-to-root plants and NAA for more recalcitrant woody plants. The efficacy of different auxins also depends on the explant type, and exposure to light. For example, a short duration (3-7 days) of dark incubation enhanced rooting in apple, and papaya. Addition of low concentration of the growth retardant daminozide to hormone-free medium enhanced post-transplanting acclimatisation of potato plants. Other supplements like activated charcoal, seaweed concentrate, pyroligneus acid, and dilution of salts in the basic MS medium can also aid root initiation and growth.

In vitro cultured papaya shoots in aerated vermiculite gave better rooting with higher post-transplanting survival than on non-aerated vermiculite and non-aerated agar. In some difficult-to-root species, e.g. apple and Saskatoon berries, in vitro rejuvenation via several subcultures is required before successful rooting of microcuttings. Individual cultivars may require adjustment to culture conditions and different number of subcultures. For example, over 90% rooting was achieved with apple microshoots, cv.

'Jonathan' by the ninth subculture, but a more recalcitrant variety 'Red Delicious' required over 30 subcultures for successful rooting. In Saskatoon berries, where rooting by traditional stem cuttings is very difficult, tissue culture rejuvenation and subsequent transplanting to the greenhouse can generate highly vigorous plants. Shoot cuttings taken from such planted hedges readily root within 2-3 weeks after a simple treatment with IBA.

Tissue culture containers provide sterile conditions and high moisture, thereby preventing contamination and desiccation of the explant, but restrict aeration and escape of metabolically generated ethylene and water vapors. Such conditions slow down growth and induce undesirable morphological changes. The ethylene build-up stunts plant growth and reduces leaf-size, and causes leaf drop. Such plants desiccate quickly upon transfer to the external environment. Vented closures, including those for baby-food-jars with built-in different sizes of microbial filters, have been used to facilitate gas exchange between culture vessels and the ambient environment.

High relative humidity in culture vessel reduces deposition of epicuticular waxes on leaves. It impairs the functioning of stomata and alters leaf mesophyll by reducing the number of palisade cells and by forming spongy parenchyma with large air spaces. Other effects include low nutrient uptake due to decreased transpiration, decreased lignification, and reduced or abnormal development of the vascular system. Although the cuticular wax development and stomata functioning can be induced by gradual adjustment to lower humidity, morphological changes cannot be easily reversed. The altered leaf morphology reduces photosynthesis and enhances respiration.

Cooling the bottom of culture vessels by 2-3°C below the ambient air temperature in the culture room or providing high light intensity can reduce high humidity. Gradual opening of vessels for a few days prior to transplanting can also adjust high humidity. Many micropropagated plants have C3 photosynthetic pathway; hence, are greatly affected in their ability to assimilate carbon. The balance between photosynthesis and photorespiration controls their productivity. Under the traditional methods of micropropagation, cultures are usually mixotrophic. The carbohydrates in the media inhibit carbon assimilation via photosynthesis by reducing rubisco (ribulose-1-5-bisphosphate carboxylase/oxygenase) activity. On the other hand, several studies indicate a stimulatory effect of sugar on biomass accumulation under elevated CO_2.

In the photoautotrophic culture systems the cultures are grown under CO_2-enrichment (1000-1500μmol mol^{-1}), increased light intensity (at least 150μmol m^{-2}, s^{-1}), good gas exchange and reduced humidity, and on media with or without or sucrose. Plantlets produced under such conditions are more vigorous, have larger root- systems, and are less susceptible to microbial contamination. They are able to photosynthesize, and do not require acclimatization during post-vitro establishment.

Priming for Temperature Tolerance

In vitro grown plantlets can be acclimatized to low temperatures for cold tolerance under field conditions. Alfalfa and red clover seedlings withstood -16 and -10°C, respectively, upon cold hardening at 2-5°C, 80μE m^{-2}, s^{-1} light and 8h photoperiod, for 3-5 weeks. Such hardening has been successfully utilized in the selection of cold tolerant alfalfa and red clover genotypes. Elevation of sucrose in the culture medium or application of abscisic acid, or both, may further improve plantlet hardiness in some species.

Acclimatization to Soil Transfer

Good quality propagules with well developed roots and leaves are easy to acclimatize to the external environment. Any successful acclimatization protocol must ensure that the plants maintain active growth during the entire weaning period. The effect of in vitro growth conditions on post-vitro acclimatization has been demonstrated in grapevine and chile ancho pepper, Capsicum annuum L.. The most common approach to improve plant survival upon transfer to soil is their gradual adaptation to the ex-vitro environment. Under such conditions, plants rapidly convert from heterotrophic or photomixotrophic to autotrophic growth, form fully functional root systems, and their stomatal and cuticular transpiration is reduced. Such gradual adaptation is often carried out in greenhouses by decreasing relative humidity using fog or mist chambers and increasing light intensity and shading techniques.

Light intensity needs to be adjusted to the plant requirements. Rooting of in vitro produced shoots in vivo is recommended for some species to promote development of roots already adapted to external conditions, to prevent potential damage to fragile in vitro roots, or to avoid post-rooting dormancy. Other treatments of in vitro plantlets upon transplanting to the outside environment include high phosphate fertilizer to enhance their vigor, fungicides to prevent damping-off and anti-transpirants to prevent desiccation; however, they may cause phytotoxicity.

A simple, low cost method for ex vitro weaning and liner production of woody plant species and Hosta has been developed. The shoots with root-initials from the rooting medium are transplanted in 'Jiffy'TM plugs. Plants are acclimatized in plastic trays (25x50x5cm) with transparent cover. Each tray holds 300 plugs, each 18mm diameter. The plugs are saturated in a 2.5-3L solution of fertilizer (1g/L) 10-52-10 NPK; the excess solution is discarded before transplanting. The plants are watered weekly (approx. 1L water/week) and kept in the growth room benches for 3-weeks after transplanting at 22-24/20°C day/night, 14h photoperiod, under 120 μEm-2s-1 fluorescent light. The current cost is US$73 for 5400 plugs, $58 for 100 trays, and $60 for 100 plastic covers; the total cost per plug being US$0.018. The rooting success in woody plants, e.g. Swedish columnar aspen, Saskatoon berry, chokecherry and pincherry cultivars, sour cherry, low bush blueberries and others was more than 95%, and 100% with hosta varieties.

In seed potato production, 2 to 4-week old tissue cultured plantlets are usually hardened off for 2-4 weeks in a greenhouse, and transplanted in the field. Such plantlets are still juvenile and not well adapted to stress. To prevent desiccation, insect and bird damage, and spread of viruses by aphid, many growers cover plants with floating covers and install an overhead irrigation system. Seed potatoes produced under floating covers are consistently better in quality than those from the unprotected planting.

In vitro produced-propagules of woody species, e.g. Saskatoon berry, Chokecherry (Prunus virginiana L.), Pincherry (P. pennsylvanica L.), Swedish columnar aspen (Populus tremula var. erecta), European weeping birch (Betula pendula var. 'Gracilis' become dormant after rooting. In vitro rooted Saskatoon berry is particularly prone to dormancy as stress factors readily induce terminal bud dormancy. When the shoots stop growth, leaf shedding occurs in the culture or soon after transplanting, and the propagules desiccate. The liners produced from such plants are less vigorous and difficult to establish in the field. In Chokecherry, dormancy symptoms of shoots with reddish leaves and cessation of terminal-bud growth are induced upon depletion of nutrients in the medium.

Post-rooting dormancy has also been reported in rhododendrons and low bush blueberries. Adding low amounts of gibberellic acid (GA) to the medium prevents dormancy, although shoots treated with GA are difficult to root (Pruski et al., 2000). In Saskatoon berry, ex-vitro rooting under non-sterile conditions is helpful, although does not eliminate the problem. Spraying leaves of ex vitro rooted-shoots with BAP (400ppm), followed by GA4+7 (100-PPM) the next day prevents dormancy.

Somatic Embryos

Somatic embryos are produced as adventitious structures directly on explants of zygotic embryos, from callus and suspension cultures. The stress inducing molecules in plants, ABA and JA have been successfully used to facilitate embryo maturation and priming for resistance to post transplanting stress. ABA and osmotic shock with 6% sucrose is a routine step to induce desiccation tolerance in alfalfa synthetic seeds. Osmotic stress induction with 3% sorbitol was effective in embryo maturation and conversion in soybean cultures and ABA treatment combined with partial desiccation of encapsulated somatic embryos in sugarcane.

Several sugar alcohols and polyethylene glycol have been tested with different degrees of success for priming SE propagules of various plants. In conifers, post ABA embryo maturation has been significantly improved by benzyl adenine (BA). The lowering of sucrose concentration from 58 to 29mM in the proliferation medium improved the formation of cotyledonary SE on the maturation medium. Interaction between plant growth regulators and various carbohydrates and organic nitrogen compounds in embryo development and maturation has been reported.

Plants in their natural environment are colonized both by external and internal microorganisms. Some microorganisms, particularly beneficial bacteria and fungi, can improve plant performance under stress environments, and consequently enhance yield. Plants infected by microorganisms develop systemic resistance (systemic acquired resistance, SAR, or inducedsystemic resistance, ISR), and/or benefit from their antagonistic abilities towards pathogens (cross protection). Although, the inoculation of seeds with beneficial microorganisms has been practiced for more than 50 years, the inoculation of tissue culture propagules to enhance plant performance is relatively new.

Plant tissue culture is based on axenic (contaminant-free) culture systems. Hence, microorganisms, including endophytes, are treated as problemcausing contaminants, and various procedures have been developed to eliminate them. Only recently, have microbial inoculants, primarily bacterial and mycorrhizal, been evaluated as propagule priming agents both as in vitro co-cultures and on transplanting. Standard microbiological techniques allow culturing only a few percent of the naturally occurring microorganisms. Moreover, the soil and plant associated bacteria can switch between culturable and unculturable stages.

In vitro Bacterization

Rhizosphere bacteria interact with plants and other inhabiting organisms by producing a wide array of chemicals. Many of these chemicals exhibit plant regulatory activity. There is increasing evidence that the bacteria associated with plants (endophytic and epiphytic) induce host resistance to environmental stresses. This is commonly linked to the production of salicilate or jasmonate, or both. These two types of signalling chemicals are associated with resistance to biotic stress. Recent molecular evidence in Arabidopsis thaliana inoculated with the plant growth promoting rhizobacterium, Paenibacillus polymyxa, indicates however, that the abiotic and biotic stress-resistance are linked.

An effective plant beneficial bacterium, Pseudomonas spp. strain PsJN, was used to enhance tolerance to transplanting stress in potato. The bacterium was originally isolated as a contaminant from Glomus vesiculiferum infected onion roots. A key element in its isolation was that the bacterium did not grow on a standard potato culture medium in the absence of plantlets. Pre-selection of plant beneficial bacteria on plant tissue culture media is thus recommended as a critical step in the development of an in vitro co-culture system, so the bacteria do not overgrow the culture.

Dipping potato nodal cuttings in a suspension of a beneficial bacterium can dramatically affect in vitro growth of plantlets. So far, our isolate, Pseudomonas sp. strain PsJN, has been the most effective plant growth promoting bacterium under in vitro conditions. It forms endophytic and epiphytic populations when co-cultured either with potato, tomato and grapevine. Thus, in clonal propagation by nodal explants, there is no need for further re-inoculation. The bacterium stimulates plant growth and induces changes that lead to better water management. It also enhances resistance to low level pathogens of potato and tomato and to Botrytis cinerea infection of grapevine.

The inoculated plantlets had a massive and wellbranched root system after four weeks in culture, and were more advanced developmentally. The stems were sturdier with more lignin deposits around the vascular system, more root hairs and more and larger leaf hairs. Stomata function of the bacterized in vitro plantlets closely resembled those of the greenhouse hardened non-bacterized transplants. In wheat seedlings exposed to osmotic stress, improvement of water relations has been observed by their co-culture with Azospirillum brasiliense strain SP245.

Enhanced transplant survival of black locust (Rabinia pseudoacacia L.), inoculated with different Rhizobium isolates selected for growth vigour from Rabinia varieties, has also been reported. The stimulation of adventitious rooting in cultures with Agrobacterium rhizogenes is well known. It was successfully used to enhance rooting of two Mexican species of Pinus. Rooting of somatic embryo-derived adventitious shoots (15-30 mm long) increased between 7-13%, and in shoots co-cultured with A. rhizogenes to 67% compared with 60% in the auxin treated controls.

Root development was also induced in slash pine explants on co-culture with the bacterium isolates. The bacterium-induced roots resembled seedling roots rather than hairy roots, which are induced with A. rhizogenes. Induction of root elongation has been observed in potato plants co-cultured in vitro with some strains of Rhizobium leguminosarum biovar trifolii, isolated from the root nodules of red clover. The bacterization of shoot explants of oregano with a Pseudomonas sp. prevented vitrification.

The oregano plantlets co-cultured with the bacterium had lower water content, but more phenolics and chlorophyll than the non-bacterized controls. Similar responses were recorded with potato and vegetable crops. Bacterized plantlets were greener, had elevated levels of cytokinins, PAL (phenylalanine ammonia-lyase), free phenolics (especially chlorogenic and caffeic acids), and contained more lignin. In potato, elevated PAL activity in bacterized plants was found during the first four weeks of growth, there being no difference in its level in six week-old plants.

However, when explants were taken from six-week-old bacterized and control stock plants, PAL activity in the newly derived shoots was higher in the pre-bacterized treatment. This suggests that pre-bacterized plantlets respond faster to cutting injuries, a critical factor in the induced host resistance to pathogen attack. In vitro bacterization of potato and vegetable plantlets significantly enhanced their post transplanting survival vigour. Enhanced vigour of potato plantlets co-cultured with isolates of Pseudomonas fluorescence isolated from tubers, during and after weaning has also been reported.

In greenhouse experiments, plants derived from dual cultures of potato and our pseudomonad bacterium had larger root systems and stolons, tuberized earlier, and gave higher tuber yields than the non-bacterized controls. The bacterized potato plantlets transplanted directly from culture vessels to the field had significantly better survival than the non-bacterized controls; however, the tuber yield varied from year to year; high precipitation or severe drought reduced the benefits of bacterization.

Tissue culture techniques provide an opportunity for the introduction of nitrogen fixing endophytes (other than Rhizobia) into clonally propagated plants for sustainable production systems. A protocol has been developed to inoculate micropropagated sugarcane plantlets with Acetobacter diazotrophicus at the end of the rooting period on medium with 10X diluted salts and sugar but without hormones and vitamins. Thirty days after transplanting, the bacteria could be re-isolated from the plant tissues. Stable artificial associations between plants and nitrogen fixing bacteria have been also reported.

A symbiotic culture system of callus and bacteria between Daucus carota L. and Azotobacter zettuovii (CRS-H6) was developed. The callus grew for four years on the nitrogen-free medium with lactose as the carbon source. The bacteria located in the intracellular spaces were transmitted to newly regenerated plantlets and fixed nitrogen. Similar stable associations were established with tomato, potato, wheat, sugarcane and poplar, using different strains of eight Azotobacter species, and in strawberries with Azomonas insignis.

In tomato, the bacterium was transmitted through the seed. Approximately, 20% of the seedlings from the seed of tomato cocultured with Azotobacter beijerinckii showed N2 fixing activity two months after germination.

Ex vitro Bacterization

To comply with the certification requirements of plant tissue culture propagules, commercial laboratories are hesitant to introduce microorganisms into their in vitro cultures. One approach is to wean plants in rhizosphere with both beneficial bacteria and mycorrhizal fungi. In potato, the tissue-cultured plantlets are readily infected both in external and internal tissues with bacteria present in the transplanting medium. This approach, although attractive, will have to deal with the problem of maintaining stable populations of the introduced microorganisms.

The beneficial effects of bacterial inoculation on tree seedlings and greenhouse produced potato tubers show the potential of post-vitro bacterization of tissue-cultured propagules. The post-vitro mycorrhization and bacterization of micropropagated strawberry, potato and azalea with certain combinations of bacteria and mycorrhiza enhanced greenhouse production of minitubers, and a mixture of three strains of rhizobacteria improved the post-transplanting performance of strawberries.

Mycorrhization

The inoculation of tissue-cultured plantlets with endomycorrhizal fungi is beneficial to plant growth. It is well known that the root colonization with vesicular-arbuscular mycorrhizae (VAM) improves plant nutritional status, water management and disease resistance. Several secondary metabolites (cyclohexenone derivatives) with structural similarity to abscisic acid have been identified in tobacco roots in response to the Glomus intraradices infection. A variety of phenylpropanoid compounds induced by ectomycorrhizal inoculants, have also been identified in several conifers. Such compounds are related to the biotic and abiotic stress management in plants.

To improve the performance of tissue-cultured plantlets, rooted-shoots have been inoculated with mycorrhiza at transplanting. The benefits of mycorrhization depended on the growing medium, plant and mycorrhizal species, and the degree of root colonization. Because VAM stimulated growth of plants was frequently expressed only after acclimatization, it was suggested to develop a culture system where the mycorrhizal fungi are introduced in vitro during the rooting stage. It is possible to grow fungi in vitro on Ri T-DNA transformed carrot roots under elevated CO_2.

Capitalizing on this technique, a 'tripartite' culture system for in vitro inoculation of strawberries and vegetable crops was developed. The most effective treatment consisted of 30 day-old VAM (Glomus intraradices) transformed carrot-root culture and strawberry shoots in cellulose plugs. After root induction, the plugs were placed on the surface of the mycorrhized root culture and kept in a growth chamber under 5000 PPM CO_2 for 20 days. All plantlets were successfully colonized, and exhibited a larger root system, better shoot growth, and a higher (more negative) osmotic potential when compared to non-mycorrhized controls. It is suggested that the enhancement of osmotic potential is important in the pre-adaptation step prior to full acclimation of plantlets for transplanting.

The high osmotic potential of wheat seedlings co-cultured with Azospirillum allowed them to withstand osmotic stress much better than non-inoculated controls with low osmotic potential. In strawberry, mycorrhized plants had a better establishment rate and produced more runners than non-mycorrhized controls. The plants were grown in polyurethane foam substrate under photoautotrophic conditions with reduced sucrose concentration in the medium. The foam tear-away strips fit into any culture vessel, and the VAM spores can be placed directly in the planting holes. In garlic, improved growth was observed after post-vitro transplant inoculation.

Increased rooting and reduction in weaning stress have been reported in the medicinal plant, Babtista tinctoria (L.) R. BR.. Several studies have demonstrated synergistic effects of VAM fungi and diazotrophic bacteria on nutrition and growth of various crops. Such effects of "mycorrhiza helper bacteria" need to be explored in tissue culture systems. Creation of defined tissue culture microecosystems could allow the study of complexed plant-microbialenvironment interactions, which could refine and improve our traditional in vitro propagation methods and prime the propagules for ex-vitro environments.

Microbial Derivatives

Cyanobacterial and fungal culture extracts have been shown to induce developmental changes and pathogen resistance in tissue-cultured material when added to the medium. Microbial culture filtrates have been reported for the control of plant tissue culture contaminants. An acetone precipitated fraction of Bacillus subtilis Ehrnberg (strain 2) and Trichoderma viride Pers. (strain A) exhibited high antifungal activity, and extracts from Pseudomonas fluorescence Migula (strain X) had high antibacterial activity. The bacterial extracts did not reduce the performance of cultures of Nicotiana tabacum L. over four subcultures but reduced the growth of accidental contaminants.

Heat-stable mycelial extracts of another Trichoderma species, a non-pathogenic T. longibrachiatum were triggered induced-resistance response to Phytophthora parasitica var.

nicotianae (race 0) in tobacco seedlings. The extract induced expression of pathogenesis-related proteins, PR-16 and osmotin (PR-5) at a higher level than the extract prepared in the same manner from the pathogenic fungus.

INCORPORATION OF LOW COST OPTIONS

Low cost options can generally be incorporated into the design of the building, laboratories, working areas, layout of equipment, lighting, and heating to provide smooth and efficient operations. However, some options are stage-specific to the process of micropropagation and others are linked to marketing. The following strategies should be considered before venturing into commercial micropropagation.

Selection of Crops

Micropropagation protocols have been developed for a large number of species; however, only a limited number of plants are being produced on a large scale through micropropagation. This is because conventional methods of propagation are cheaper than tissue culture technology or there is a selective market demand. Ornamentals account for approximately 80% of the world trade in the tissue culture industry. Plants produced by commercial laboratories should yield steady revenue, e.g. Spathiphyllum, Syngonium, lilies. Some plants are unique to individual commercial companies.

The life cycle of the species and the cropping system adopted by growers are important criteria to select plants for large-scale tissue culture. Initially, micropropagation units focused on the production of ornamental plants that were least affected by the season, e.g. foliage and indoor plants. But to capture larger markets, commercial tissue culture units have ventured into fruits, vegetables, landscape and medicinal plants, which provide options for production around the year. This leads to improved cash flow and optimal use of equipment and facilities. When selecting specific varieties protected under Plant Breeders Rights, these should be multiplied only with the agreement of the breeders.

Prevention of Variation in Tissue Culture

The establishment of mother cultures from explants is a significant part of the production cost. Production projections can get out of gear due to the lack of sufficient mother cultures and maintenance of large numbers of such cultures adds to the cost of production. There is a general consensus that the number of subcultures for explants should be limited to avoid somaclonal variation. For example, in potato, the number of subcultures is recommended to be around twenty. Hence, repeated explant culture initiation becomes necessary to have sufficient number of mother cultures whenever largescale production is planned.

For many plants, it is not known whether limiting the number of subcultures reduces genetic variation. Many of the plants that are propagated from vegetative parts such as banana, potato, sweet potato, sugarcane, chrysanthemum and many fruits are polyploids and well buffered against the expression of mutations and somaclonal variation. If axillary bud proliferation is used as the strategy of multiplication, the frequency of variation is low. If variation does occur, the number of variants will rise with increasing number of subcultures unless there is selective elimination in tissue culture. The high frequency of somaclonal variation has prevented the use of tissue culture for mass scale propagation of oil palm. In general, the initiation of cultures from meristem-tip gives low variation in subcultures. Some of the novel somaclonal variants can be of value in expanding the market.

Multiplication and Hardening

Liquid media reduce the costs of multiplication. Increasing the multiplication rate through enhanced axillary bud proliferation helps in cutting cost of production. Mechanical cutting and bioreactors is a valuable option at Stage II. In some systems, shoot elongation and rooting are two separate phases. Combining them in one step by using appropriate media with auxins and cytokinins is a valid cost cutting strategy. Reducing the duration in the hardening facility but increasing the survival rate reduces production costs. If natural light is used during multiplication and rooting, the plants need relatively shorter duration for hardening.

Acclimatization in vitro helps in better survival when the plants are transferred to soil. Loss of plants at the hardening stage contributes to increased costs of production. A cost-effective system based on natural sunlight for in vitro acclimatization gives high survival. If grown under artificial lighting, tissue cultured plants should be kept under shade for some time and only then subjected to full sun hardening.

Greenhouse Facility

Sophisticated and costly greenhouses can be substituted by cost-effective techniques for hardening in vitro derived plantlets. For this, it is necessary to understand the changes that tissue-cultured plantlets go through during the transfer from in vitro to in vivo conditions. Micropropagated plantlets should be acclimatized gradually by transferring them to a clean area under partial shade.

In a low-cost hardening system, the culture containers are kept in the greenhouse with lids loosened. The plants are allowed to grow for 4-6 weeks under conventional Grow lights TM. This semi-hardens plants, and leads to shoot elongation, which is usually done under laboratory conditions. Contamination can be controlled with regular sprays of fungicides and bactericides. The plants are periodically sprayed with water to maintain humidity. Bench lamps are used to provide light, if and when required.

Semihardened elongated shoot clumps are then separated under non-sterile conditions, and rooted in peat or plugs under thin plastic sheets for 2-3 weeks. This low-cost alternative reduces the capital cost of hardening by saving laboratory space, labour, and time. This procedure has been used on a commercial scale for a number of plants such as Gerbera, Spathiphyllum, Syngonium, African violets, Chrysanthemum, Ficus, Philodendron and other ornamentals in the Netherlands and Israel. The growers in Florida harden tissue-cultured foliage plants in cavity trays covered with a transparent plastic tray, thus saving costs of a greenhouse. In some parts of India, hanging the culture plastic bags in shade hardens tissue-cultured sugarcane plants.

Plants produced in this manner are semihardened, and can be delivered to the growers at a relatively cheaper price. Low-cost plastic tunnels (polytunnel) with bio-fertilization can also be used for hardening, especially in cooler climates. A rectangular pit of desired and manageable size is dug on a site free of water logging. A frame of bamboo or any other available material is made above the pit with a gentle slope to one side. The frame is then covered with a clear polythene sheet, and tied using nylon ropes. The sheet is sealed with mud on three sides, leaving one side free.

The temperature and humidity inside these structures is generally higher than the outside, and protects the plants from frost. The polythene sheet can be partly opened to allow air circulation and sunlight as and when required. The added advantage of polypits is the carbon dioxide fertilization effect and reduced need for watering. The International Institute for Tropical Agriculture, Ibadan, Nigeria has developed another low cost hardening alternative. The institute provides the farmers with plants in plastic bags and handling instructions. For hardening the plants, the farmers hang the plastic bags under tree shade.

CHAPTER 4

PLANT TISSUE CULTURE

Plant tissue culture refers to growing and multiplication of cells, tissues and organs on defined solid or liquid media under aseptic and controlled environment. Plant tissue culture technology is being widely used for large-scale plant multiplication. The commercial technology is primarily based on micropropagation, in which rapid proliferation is achieved from tiny stem cuttings, axillary buds, and to a limited extent from somatic embryos, cell clumps in suspension cultures and bioreactors.

Plant tissue cultures are initiated from tiny pieces, called explants, taken from any part of a plant. Practically all parts of a plant have been used successfully as a source of explants. In practice, the "explant" is removed surgically, surface sterilized and placed on a nutrient medium to initiate the mother culture, that is multiplied repeatedly by subculture. The following plant parts are extensively used in commercial micropropagation.

Shoot-tip and meristem-tip culture: Shoots develop from a small group of cells known as shoot apical meristem. The apical meristem maintains itself, gives rise to new tissues and organs, and communicates signals to the rest of the plant. Shoot-tips and meristem-tips are perhaps the most popular source of explants to initiate tissue cultures. The shoot apex explant measures between 100 to 500µm and includes the apical meristem with 1 to 3 leaf primordia. The apical meristem of a shoot is the portion lying distal to the youngest leaf primordium, and is ca.100μm in diameter and 250μm in length with 800-1200 cells.

In practice, shoot-tip explants between 100 to 1000μm are cultured to free plants from viruses. Even explants larger than 1000μm have been frequently used. The term "meristem-tip culture" has been suggested to distinguish the large explants from those used in conventional propagation.

Nodal or axillary bud culture: This consists of a piece of stem with axillary bud culture with or without a portion of shoot. When only the axillary bud is taken, it is designated as "axillary bud" culture.

Floral meristem and bud culture: Such explants are not commonly used in commercial propagation, but floral meristems and buds can generate complete plants. Other sources of explants: In some plants, leaf discs, intercalary meristems from nodes, small pieces of stems, immature zygotic embryos and nucellus have also been used as explants to initiate cultures.

Cell suspension and callus cultures: Plant parts such as leaf discs, intercalary meristems, - stem-pieces, immature embryos, anthers, pollen, microspores and ovules have been cultured to initiate callus. A callus is a mass of unorganized cells, which in many cases, upon transfer to suitable medium, is capable of giving rise to shoot-buds and somatic embryos, which then form complete plants. Such calli on culture in liquid media on shakers are used for initiating cell suspensions. Liquid suspension cultures maintained on mechanical shakers achieve fast and excellent multiplication rates. However, in commercial micropropagation, calli are cultured mostly in bottles and flasks kept on semi-solid or liquid media. To a limited extent, bioreactors have become popular for somatic embryogenic cultures. It is considered that some day robotics could be adapted to bioreactor-based micropropagation.

Cultured Cells and Tissue

The cultured cells and tissue can take several pathways to produce a complete plant. Among these, the pathways that lead to the production of true-to-type plants in large numbers are the popular and preferred ones for commercial multiplication. The following terms have been used to describe various pathways of cells and tissue in culture.

Regeneration and organogenesis

In this pathway, groups of cells of the apical meristem in the shoot apex, axillary buds, root tips, and floral buds are stimulated to differentiate and grow into shoots and ultimately into complete plants. In many cases, the axillary buds formed in the culture undergo repetitive proliferation, and produce large number of tiny plants. The plants are then separated from each other and rooted either in the next stages of micropropagation or in vivo (in trays, small pots or beds in glasshouse or plastic tunnel under relatively high humidity). The explants cultured on relatively high amounts of auxin (e.g. (2,4-D, 2,4-dichlorophenoxyacetic acid) form an unorganized mass of cells, called callus. The callus can be further sub-cultured and multiplied. The callus shaken in a liquid medium produces cell suspension, which can be subcultured and multiplied into more liquid cultures.

The cell suspensions form cell clumps, which eventually form calli and give rise to plants through organogenesis or somatic embryogenesis. In some cases, explants e.g. leaf-discs and epidermal tissue can also generate plants by direct organogenesis and somatic embryogenesis without intervening callus formation, e.g. in orchardgrass. Dactylis glomerata L.. In organogenesis the cultured plant cells and cell clumps (callus) and mature differentiated cells (microspores, ovules) and tissues (leaf discs, inter-nodal segments) are induced to differentiate into complete plants to form shoot buds and eventually shoots, and rooted to form complete plants.

Somatic embryogenesis

In this pathway, cells or callus cultures on solid media or in suspension cultures form embryo-like structures called somatic embryos, which on germination produce complete plants. The primary somatic embryos are also capable of producing more embryos through secondary somatic embryogenesis. Although, somatic embryogenesis has been demonstrated in a very large number of plants and trees, the use of somatic embryos in large-scale commercial production has been restricted to only a few plants, such as carrot, date palm, and a few forest trees.

Somatic embryos are produced as adventitious structures directly on explants of zygotic embryos, from callus and suspension cultures. Somatic embryos and synthetic seeds (embryos encapsulated in artificial endosperm) hold potential for large-scale clonal propagation of superior genotypes of heterogeneous plants. They have also been used in commercial plant production and for the multiplication of parental genotypes in large-scale hybrid seed production. In many species, somatic embryos are morphologically similar to the zygotic embryos, although some biochemical, physiological and anatomical differences have been documented.

The synthetic auxin, 2,4-D is commonly used for embryo induction. In many angiosperms, e.g., carrot and alfalfa, subculture of cells from 2,4-D containing medium to auxin-free medium is sufficient to induce somatic embryogenesis. The process can be enhanced with the application of osmotic stress, manipulation of medium nutrients, and reducing humidity. Selection of embryogenic cell lines has also been successfully used. For example, selection for unique morphotypes in grapevine cultures allows production of high quality embryos with predicable frequency.

A major problem in large-scale production of somatic embryos is culture synchronization. This is achieved through selecting cells or pre-embryonic cell clusters of certain size, and manipulation of light and temperature, temporary starvation or by adding cell cycle synchrönizing chemicals to the medium. Cytokinins seem to play a key role in cell cycle synchronization and embryo induction, proliferation and differentiation. Abscisic acid is crucial in all the stages of somatic development, maturation and hardening.

Synthetic seeds

The concept of production and utilization of synthetic seeds (somatic embryo as substitutes for true seeds) was first suggested by Murashige in 1977. Synthetic seeds can be produced either as coated or non-coated, desiccated somatic embryos or as embryos encapsulated in hydrated gel (usually calcium alginate). Successful utilization of synthetic seeds as propagules of choice requires an efficient and reproducible production system and a high percentage of post-planting conversion into vigorous plants. Artificial coats and gel capsules containing nutrients, pesticides and beneficial organisms have long been thought as substitutes for seed coat and endosperm. However, this technology is still in the developmental stage, and currently cannot compete with the other methods of commercial plant propagation.

MICROPROPAGATION

The process of plant micropropagation aims to produce clones (true copies of a plant in large numbers). The process is usually divided into the following stages:

Stage 0- pre-propagation step or selection and pre-treatment of suitable plants.

Stage I - initiation of explants - surface sterilization, establishment of mother explants.

Stage II - subculture for multiplication/proliferation of explants.

Stage III - shooting and rooting of the explants.

Stage IV - weaning/hardening.

These stages are universally applicable in large-scale multiplication of plants. The individual plant species, varieties and clones require specific modification of the growth media, weaning and hardening conditions. A rule of the thumb is to propagate plants under conditions as natural or similar to those in which the plants will be ultimately grown ex-vitro. For example, if a chrysanthemum variety is to be grown under long day-length for flower production, it is better to multiply the material under long-day length at stages III and IV. There is a wide option to undertake production of plant material up to a limited number of stages. For example, many commercial tissue culture companies undertake production up to Stage III, and leave the remaining stages to others.

Stage 0

The stage 0 also pre-propagation stage requires proper maintenance of the mother plants in the greenhouse under disease- and insect-free conditions with minimal dust. Clean enclosed areas, glasshouses, plastic tunnels, and net-covered tunnels, provide high quality explant source plants with minimal infection. Collection of plant material for clonal propagation should be done after appropriate pretreatment of the mother plants with fungicides and pesticides to minimize contamination in the in vitro cultures. This improves growth and multiplication rates of in vitro cultures. The control of contamination begins with the pretreatment of the donor plants. They may be prescreened for diseases, isolated and treated to reduce contamination.

The explants are then brought to the production facility, surface sterilized and introduced into culture. They may at this stage be treated with antibiotics and fungicides as well as anti-microbial formulations, such as PPM. The explants are then culture indexed for contamination by standard microbiological techniques, which are occasionally supplemented with tests based on molecular biology or other techniques.

Stage I

This stage refers to the inoculation of the explants on sterile medium to initiate aseptic culture. Initiation of explants is the very first step in micropropagation. A good clean explant, once established in an aseptic condition, can be multiplied several times; hence, explant initiation in an aseptic condition should be regarded as a critical step in micropropagation. More than often, explants fail to establish and grow, not due to the lack of a suitable medium but because of contamination. The explants are transferred to in vitro environment, free from microbial contaminants. The process requires excision of tiny plant pieces and their surface sterilization with chemicals such as sodium hypochlorite, ethyl alcohol and repeated washing with sterile distilled water before and after treatment with chemicals.

After a short period of culture, usually 3 to 5 days, the contaminated explants are discarded. The surviving explants showing growth are maintained and used for further subculture. In herbaceous plants e.g. potato, chrysanthemum, carnation, streptocarpus, strawberry, and African violet; the explant sources are meristems, apical- and axillary buds, young seedlings, developing young leaves and petioles, and unopened floral buds. The following low cost options can be adapted to initiate explants:

Sterile instrument technique

This method assumes that most of the deep-seated meristems and those covered by leaves or other integuments (e.g. floral bracts) are sterile. In this procedure, the explant is washed with sterile water, rinsed in ethanol, and instruments are sterilized every time they touch the surface of the explant, and the explant is moved to a new location on the dissection stage.

Surface sterilization technique

This is by far the most commonly used method. The explants are washed in sterile water, rinsed in ethanol, and surface sterilization is achieved by using chemicals with chlorine base. Calcium or sodium hypochlorite based solutions, 1-3% (v/v) are usually used for soft herbaceous materials. A cheap and ready-made sterilant is 5-7% solution of 'Domestos'- a toilet disinfectant which contains 10.5% v/v sodium hypochlorite, 0.3% sodium carbonate, 10.0% sodium chloride and 0.5% (w/v) sodium hydroxide and a patented thickener. The explants are washed in sterile distilled water before and after sterilization. Other surface sterilants used include mercuric chloride (avoid its use as far as possible, since it is highly toxic), hydrogen peroxide, and potassium permanganate. The following protocols are available:

For soft tissues

1. Wash explants from perennial plants for 1-2 hr in tap water. Eliminate this step for material from glasshouse grown plants.
2. Wash in sterile distilled water three to four times for 5 to 10 minutes each.
3. Dip in 95% ethanol for 3 to 5 seconds.
4. Wash once again with sterile distilled water for 5 minutes.
5. Surface-sterilize in 5% 'Domestos' (v/v) for 20-25 minutes.
6. Wash with sterile distilled water three times for 10 minutes each.
7. Drain water droplets by placing on pre-sterilized blotting paper.
8. Transfer explants singly to the medium.

For woody stems (e.g. roses, hardy shrubs, and trees):

1. Collect stems, shoots, buds and store at 5 0C till needed.
2. Rinse in ethanol for 3 to 5 seconds.

3. Rinse in 1-% sodium hypochlorite (20% bleach) for 10 minutes.
4. Place lower parts of stems in flasks in 2% sucrose and 200 PPM 8-hydroxyquinoline citrate at 23+2 0C. For items collected in September/October, add 50-PPM GA3. After that 10 PPM GA will help break the dormancy.
5. Re-cut the bottom of stem and replace the solution after 2 days.
6. Excise the softwood from the developed shoots and use material for explants or for rooting.
7. Surface-sterilize as in the above protocol.

Do not forget to sterilize forceps and scalpel every time for the transfer of explants to fresh solutions. Use sterile containers in the protocol of surface sterilization. If explants become brown or pale at the end of the protocol, reduce the strength of 'Domestos' to 2.5%. Alternatively, dip explants in 10% 'Domestos' for 2 minutes and then proceed to surface sterilize with 3-5% 'Domestos' for 20 minutes. If basal contamination is observed after 2-3 days of culture, explants can sometimes be rescued by removing the basal end by making a single cut with a sharp scalpel and re-culturing on fresh medium.

Stage II

Stage II is the propagation phase in which the explants are cultured on the appropriate media for multiplication of shoots. The primary goal is to achieve propagation without losing the genetic stability. Repeated culture of axillary and adventitious shoots, cutting with nodes, somatic embryos and other organs from Stage I leads to multiplication of propagules in large numbers. The propagules produced at this stage can be further used for multiplication by their repeated culture. Sometimes it is necessary to subculture the in vitro derived shoots onto different media for elongation.

Stage III

The in vitro shoots obtained at Stage II are rooted to produce complete plants. If the proliferated material consists of bud-like structures (e.g. orchids) or clumps of shoots (banana, pineapple), they should be separated after rooting and not before. Many plants (e.g. banana, pineapple, roses, potato, chrysanthemum, strawberry, mint, several grasses and many more) can be rooted on half-strength-MS medium without any growth-regulators. Good sturdy well-rooted plants are essential for high survival during weaning and later transfer to soil. This stage is labour intensive and expensive. The process of in vitro rooting has been estimated to account for approximately 35-75% of the total cost of production. Efforts should be made to combine rooting and acclimatization stages.

Stage IV

At this stage, the in vitro micropropagated plants are weaned and hardened. This is the final stage of the tissue culture operation after which the micropropagated plantlets are ready for transfer to the greenhouse. Steps are taken to grow individual plantlets capable of carrying out photosynthesis. The hardening of the tissue-cultured plantlets is done gradually from high to low humidity and from low light intensity to high intensity conditions. If grown on solid medium, most of the agar can be removed gently by rinsing with water.

Plants can be left in shade for 3 to 6 days where diffused natural light conditions them to the new environment. The plants are then transferred to an appropriate substrate (sand, peat, compost, etc.), and gradually hardened. Low-cost options include the use of plastic domes or tunnels, which reduces the natural light intensity and maintains high relative humidity during the hardening process. If the plants are still joined together after rooting, these should be planted as bunches in the soil and separated after 6 to 8 weeks of growth.

Plasticity and Totipotency

Plasticity and totipotency, are central to understanding plant cell culture and regeneration. Plants, due to their sessile nature and long life span, have developed a greater ability to endure extreme conditions and predation than have animals. Many of the processes involved in plant growth and development adapt to environmental conditions. This plasticity allows plants to alter their metabolism, growth and development to best suit their environment. Particularly important aspects of this adaptation, as far as plant tissue culture and regeneration are concerned, are the abilities to initiate cell division from almost any tissue of the plant and to regenerate lost organs or undergo different developmental pathways in response to particular stimuli.

When plant cells and tissues are cultured in vitro they generally exhibit a very high degree of plasticity, which allows one type of tissue or organ to be initiated from another type. In this way, whole plants can be subsequently regenerated. This regeneration of whole organisms depends upon the concept that all plant cells can, given the correct stimuli, express the total genetic potential of the parent plant. This maintenance of genetic potential is called 'totipotency'. Plant cell culture and regeneration do, in fact, provide the most compelling evidence for totipotency. In practical terms though, identifying the culture conditions and stimuli required to manifest this totipotency can be extremely difficult and it is still a largely empirical process.

When cultured in vitro, all the needs, both chemical (Table 1) and physical, of the plant cells have to met by the culture vessel, the growth medium and the external environment. The growth medium has to supply all the essential mineral ions required for growth and development. In many cases, it must also supply additional organic supplements such as amino acids and vitamins. Many plant cell cultures, as they are not photosynthetic, also require the addition of a fixed carbon source in the form of a sugar. One other vital component that must also be supplied is water, the principal biological solvent. Physical factors, such as temperature, pH, the gaseous environment, light and osmotic pressure, also have to be maintained within acceptable limits.

PLANT CELL CULTURE MEDIA

Culture media used for the in vitro cultivation of plant cells are composed of three basic components:

— essential elements, or mineral ions, supplied as a complex mixture of salts;

— an organic supplement supplying vitamins and/or amino acids; and

— a source of fixed carbon; usually supplied as the sugar sucrose.

For practical purposes, the essential elements are further divided into the following categories:

— macroelements (or macronutrients);

— microelements (or micronutrients); and

— an iron source.

Table 1: Some of the elements important for plant nutrition and their physiological function.

Element	Function
Nitrogen	Component of proteins, nucleic acids and some coenzymes Element required in greatest amount
Potassium	Regulates osmotic potential, principal inorganic cation
Calcium	Cell wall synthesis, membrane function, cell signalling
Magnesium	Enzyme cofactor, component of chlorophyll
Phosphorus	Component of nucleic acids, energy transfer, component of intermediates in respiration and photosynthesis

Sulphur	Component of some amino acids (methionine, cysteine) and some cofactors
Chlorine	Required for photosynthesis
Iron	Electron transfer as a component of cytochromes
Manganese	Enzyme cofactor
Cobalt	Component of some vitamins
Copper	Enzyme cofactor, electron-transfer reactions
Zinc	Enzyme cofactor, chlorophyll biosynthesis
Molybdenum	Enzyme cofactor, component of nitrate reductase

Complete, plant cell culture medium is usually made by combining several different components, as outlined in Table 2.

Components of Culture Media

Macroelements: The stock solution supplies those elements required in large amounts for plant growth and development, as is implied by the name. Nitrogen, phosphorus, potassium, magnesium, calcium and sulphur are usually regarded as macroelements. These elements usually comprise at least 0.1% of the dry weight of plants. Nitrogen is most commonly supplied as a mixture of nitrate ions, from the KNO_3, and ammonium ions, from the NH_4NO_3. Theoretically, there is an advantage in supplying nitrogen in the form of ammonium ions, as nitrogen must be in the reduced form to be incorporated into macromolecules. Nitrate ions therefore need to be reduced before incorporation. However, at high concentrations, ammonium ions can be toxic to plant cell cultures and uptake of ammonium ions from the medium causes acidification of the medium.

In order to use ammonium ions as the sole nitrogen source, the medium needs to be buffered. High concentrations of ammonium ions can also cause culture problems by increasing the frequency of vitrification. Using a mixture of nitrate and ammonium ions has the advantage of weakly buffering the medium as the uptake of nitrate ions causes OH- ions to be excreted. Phosphorus is usually supplied as the phosphate ion of ammonium, sodium or potassium salts. High concentrations of phosphate can lead to the precipitation of medium elements as insoluble phosphates.

Table 2: Composition of a typical plant culture medium. The medium described here is that of Murashige and Skoog (MS)

Essential element	Concentration in stock solution (mg/l)	Concentration in medium (mg/l)
Macroelementsb		
NH_4NO_3	33000	1,650
KNO_3	38000	1 900
$CaCl_2.2H_2O$	8800	440
$MgSO_4.7H_2O$	7400	370
KH_2PO_4	3400	170
Microelements		
KI	166	0.83
H_3BO_3	1240	6.2
$MnSO_4.4H_2O$	4460	22.3
$ZnSO_4.7H_2O$	1720	8.6
$Na_2MoO_4.2H_2O$	50	0.25
$CuSO_4.5H_2O$	5	0.025
$CoCl_2.6H_2O$	5	0.025
Iron sourcec		
$FeSO_4.7H_2O$	5560	27.8
$Na_2EDTA.2H_2O$	7460	37.3
Organic supplement		
Myoinositol	20000	100
Nicotinic acid	100	0.5
Pyridoxine-HCl	100	0.5

Thiamine-HCl	100	0.5
Glycine	400	2
Carbon source		
Sucrose	Added as solid	30 000

Microelements: These elements are required in trace amounts for plant growth and development, and have many and diverse roles. Manganese, iodine, copper, cobalt, boron, molybdenum, iron and zinc usually comprise the microelements, although other elements such as nickel and aluminium are frequently found in some formulations. Iron is usually added as iron sulphate, although iron citrate can also be used. Ethylenediaminetetraacetic acid (EDTA) is usually used in conjunction with the iron sulphate. The EDTA complexes with the iron so as to allow the slow and continuous release of iron into the medium. Uncomplexed iron can precipitate out of the medium as ferric oxide.

Organic supplements: Only two vitamins, thiamine and myoinositol are considered essential for the culture of plant cells in vitro. However, other vitamins are often added to plant cell culture media for historical reasons. Amino acids are also commonly included in the organic supplement. The most frequently used is glycine (arginine, asparagine, aspartic acid, alanine, glutamic acid, glutamine and proline are also used), but in many cases its inclusion is not essential. Amino acids provide a source of reduced nitrogen and, like ammonium ions, uptake causes acidification of the medium. Casein hydrolysate can be used as a relatively cheap source of a mix of amino acids.

Carbon source: Sucrose is cheap, easily available, readily assimilated and relatively stable and is therefore the most commonly used carbon source. Other carbohydrates such as glucose, maltose, galactose and sorbitol can also be used, and in specialised circumstances may prove superior to sucrose.

Gelling agents: Media for plant cell culture in vitro can be used in either liquid or 'solid' forms, depending on the type of culture being grown. For any culture types that require the plant cells or tissues to be grown on the surface of the medium, it must be solidified. Agar, produced from seaweed, is the most common type of gelling agent, and is ideal for routine applications. However, because it is a natural product, the agar quality can vary from supplier to supplier and from batch to batch. For more demanding applications, a range of purer gelling agents are available. Purified agar or agarose can be used, as can a variety of gellan gums.

PLANT GROWTH REGULATORS

The essential point as far as plant cell culture is concerned is that, due to plasticity and totipotency, specific media manipulations can be used to direct the development of plant cells in culture. Plant growth regulators are the critical media components in determining the developmental pathway of the plant cells. The plant growth regulators used most commonly are plant hormones or their synthetic analogues. There are five main classes of plant growth regulator used in plant cell culture, namely:

— auxins;

— cytokinins;

— gibberellins;

— abscisic acid;

— ethylene.

Auxins: Auxins promote both cell division and cell growth The most important naturally occurring auxin is IAA, but its use in plant cell culture media is limited because it is unstable to both heat and light. Occasionally, amino acid conjugates of IAA, such as indole-acetyl-alanine and indole-acetyl-glycine, which are more stable, are used to partially alleviate the problems associated with the use of IAA. It is more common, though, to use stable chemical analogues of IAA as a source of auxin in plant cell culture media. 2,4-Dichlorophenoxyacetic acid is the most commonly used auxin and is extremely effective in most circumstances. Other auxins are available (Table 3), and some may be more effective or 'potent' than 2,4-D in some instances.

Cytokinins: Cytokinins promote cell division. Naturally occurring cytokinins are a large group of structurally related compounds. Of the naturally occurring cytokinins, two have some use in plant tissue culture media (Table 4). These are zeatin and 2iP (2-isopentyl adenine). Their use is not widespread as they are expensive and relatively unstable. The synthetic analogues, kinetin and BAP (benzylaminopurine), are therefore used more frequently. Non-purine-based chemicals, such as substituted phenylureas, are also used as cytokinins in plant cell culture media. These substituted phenylureas can also substitute for auxin in some culture systems.

Gibberellins: There are numerous, naturally occurring, structurally related compounds termed 'gibberellins'. They are involved in regulating cell elongation, and are agronomically important in determining plant height and fruit-set. Only a fewof the gibberellins are used in plant tissue culture media, GA3 being the most common.

Table 3: Commonly used auxins

Abbreviation/name	Chemical name
2,4-D	2,4-dichlorophenoxyacetic acid
2,4,5-T	2,4,5-trichlorophenoxyacetic acid
Dicamba	2-methoxy-3,6-dichlorobenzoic acid
IAA	Indole-3-acetic acid
IBA	Indole-3-butyric acid
MCPA	2-methyl-4-chlorophenoxyacetic acid
NAA	1-naphthylacetic acid
NOA	2-naphthyloxyacetic acid
Picloram	4-amino-2,5,6-trichloropicolinic acid

Abscisic acid: Abscisic acid (ABA) inhibits cell division. It is most commonly used in plant tissue culture to promote distinct developmental pathways such as somatic.

Ethylene: Ethylene is a gaseous, naturally occurring, plant growth regulator most commonly associated with controlling fruit ripening in climacteric fruits, and its use in plant tissue culture is not widespread. It does, though, present a particular problem for plant tissue culture. Some plant cell cultures produce ethylene, which, if it builds up sufficiently, can inhibit the growth and development of the culture. The type of culture vessel used and its means of closure affect the gaseous exchange between the culture vessel and the outside atmosphere and thus the levels of ethylene present in the culture.

Table 4: Commonly used cytokinins

Abbreviation/name	Chemical name
AP	6-benzylaminopurine
2iP (IPA)	[N^6-(2-isopentyl)adenine]
Kinetin	6-furfurylaminopurine
Thidiazuron	1-phenyl-3-(1,2,3-thiadiazol-5-yl)urea
Zeatin	4-hydroxy-3-methyl-trans-2-butenylaminopurine

Tissue Culture and Plant Growth Regulators

Generalisations about plant growth regulators and their use in plant cell culture media have been developed from initial observations made in the 1950s. There is, however, some considerable difficulty in predicting the effects of plant growth regulators: this is because of the great differences in culture response between species, cultivars and even plants of the same cultivar grown under different conditions. However, some principles do hold true and have become the paradigm on which most plant tissue culture regimes are based.

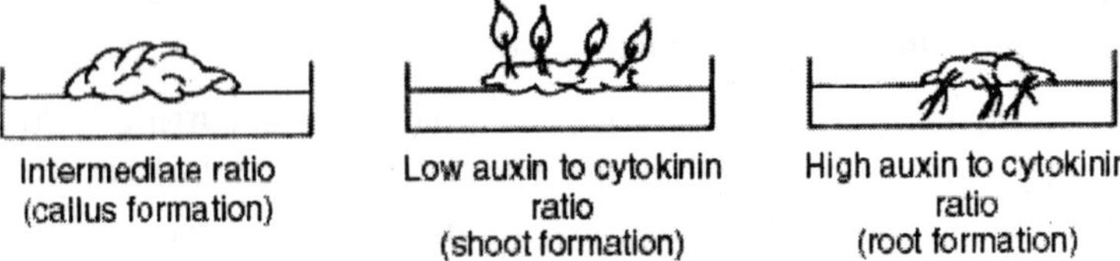

Figure 1: The effect of different ratios of auxin to cytokinin on the growth and morphogenesis of callus.

Auxins and cytokinins are the most widely used plant growth regulators in plant tissue culture and are usually used together, the ratio of the auxin to the cytokinin determining the type of culture established or regenerated (Figure1). A high auxin to cytokinin ratio generally favours root formation, whereas a high cytokinin to auxin ratio favours shoot formation. An intermediate ratio favours callus production.

TYPES OF CULTURE

Cultures are generally initiated from sterile pieces of a whole plant. These pieces are termed 'explants', and may consist of pieces of organs, such as leaves or roots, or may be specific cell types, such as pollen or endosperm. Many features of the explant are known to affect the efficiency of culture initiation. Generally, younger, more rapidly growing tissue is most effective. Several different culture types most commonly used in plant transformation studies will now be examined in more detail.

Callus Culture

Callus cultures are extremely important in plant biotechnology. Explants, when cultured on the appropriate medium, usually with both an auxin and a cytokinin, can give rise to an unorganised, growing and dividing mass of cells. It is thought that any plant tissue can be used as an explant, if the correct conditions are found. In culture, this proliferation can be maintained more or less indefinitely, provided that the callus is subcultured on to fresh medium periodically. Callus is usually composed of unspecialised parenchyma cells.

During callus formation there is some degree of dedifferentiation, both in morphology and metabolism. One major consequence of this dedifferentiation is that most plant cultures lose the ability to photosynthesise. This has important consequences for the culture of callus tissue, as the metabolic profile will probably not match that of the donor plant. This necessitates the addition of other components-such as vitamins and, most importantly, a carbon source-to the culture medium, in addition to the usual mineral nutrients. Callus culture is often performed in the dark as light can encourage differentiation of the callus.

During long-term culture, the culture may lose the requirement for auxin and/or cytokinin. This process, known as 'habituation', is common in callus cultures from some plant species, such as sugar beet. Manipulation of the auxin to cytokinin ratio in the medium can lead to the development of shoots, roots or somatic embryos from which whole plants can subsequently be produced. Callus cultures can also be used to initiate cell suspensions, which are used in a variety of ways in plant transformation studies.

Cell-Suspension Cultures

Callus cultures, broadly speaking, fall into one of two categories: compact or friable. In compact callus the cells are densely aggregated, whereas in friable callus the cells are only loosely associated with each other and the callus becomes soft and breaks apart easily. Friable callus provides the inoculum to form cell-suspension cultures. Explants from some plant species or particular cell types tend not to form friable callus, making cell-suspension initiation a difficult task. The friability of callus can sometimes be improved by manipulating the medium components or by repeated subculturing. The friability of the callus can also sometimes be improved by culturing it on 'semi-solid' medium.

When friable callus is placed into a liquid medium and then agitated, single cells and/or small clumps of cells are released into the medium. Under the correct conditions, these released cells continue to grow and divide, eventually producing a cell-suspension culture. A relatively large inoculum should be used when initiating cell suspensions so that the released cell numbers build up quickly. The inoculum should not be too large though, as toxic products released from damaged or stressed cells can build up to lethal levels. Large cell clumps can be removed during subculture of the cell suspension. Cell suspensions can be maintained relatively simply as batch cultures in conical flasks.

They are continually cultured by repeated subculturing into fresh medium. This results in dilution of the suspension and the initiation of another batch growth cycle. The degree of dilution during subculture should be determined empirically for each culture. Too great a degree of dilution will result in a greatly extended lag period or, in extreme cases, death of the transferred cells. After subculture, the cells divide and the biomass of the culture increases in a characteristic fashion, until nutrients in the medium are exhausted and/or toxic by-products build up to inhibitory levels-this is called the 'stationary phase'. If cells are left in the stationary phase for too long, they will die and the culture will be lost. Therefore, cells should be transferred as they enter the stationary phase. It is therefore important that the batch growth-cycle parameters are determined for each cell-suspension culture.

Protoplasts

Protoplasts are plant cells with the cell wall removed. Protoplasts are most commonly isolated from either leaf mesophyll cells or cell suspensions, although other sources can be used to advantage. Two general approaches to removing the cell wall can be taken-mechanical or enzymatic isolation. Mechanical isolation, although possible, often results in low yields, poor quality and poor performance in culture due to substances released from damaged cells. Enzymatic isolation is usually carried out in a simple salt solution with a high osmoticum, plus the cell wall degrading enzymes. It is usual to use a mix of both cellulase and pectinase enzymes, which must be of high quality and purity.

Protoplasts are fragile and easily damaged, and therefore must be cultured carefully. Liquid medium is not agitated and a high osmotic potential is maintained, at least in the initial stages. The liquid medium must be shallow enough to allow aeration in the absence of agitation. Protoplasts can be plated out on to solid medium and callus produced. Whole plants can be regenerated by organogenesis or somatic embryogenesis from this callus. Protoplasts are ideal targets for transformation by a variety of means.

Root Cultures

Root cultures can be established in vitro from explants of the root tip of either primary or lateral roots and can be cultured on fairly simple media. The growth of roots in vitro is potentially unlimited, as roots are indeterminate organs. Although the establishment of root cultures was one of the first achievements of modern plant tissue culture, they are not widely used in plant transformation studies.

Meristem Culture and Shoot Tip

The tips of shoots, which contain the shoot apical meristem, can be cultured in vitro, producing clumps of shoots from either axillary or adventitious buds. This method can be used for clonal propagation. Shoot meristem cultures are potential alternatives to the more commonly used methods for cereal regeneration as they are less genotype-dependent and more efficient

Embryo Culture

Embryos can be used as explants to generate callus cultures or somatic embryos. Both immature and mature embryos can be used as explants. Immature, embryo-derived embryogenic callus is the most popular method of monocot plant regeneration.

Microspore Culture

Haploid tissue can be cultured in vitro by using pollen or anthers as an explant. Pollen contains the male gametophyte, which is termed the 'microspore'. Both callus and embryos can be produced from pollen. Two main approaches can be taken to produce in vitro cultures from haploid tissue. The first method depends on using the anther as the explant. Anthers can be cultured on solid medium. Pollen-derived embryos are subsequently produced via dehiscence of the mature anthers. The dehiscence of the anther depends both on its isolation at the correct stage and on the correct culture conditions. In some species, the reliance on natural dehiscence can be circumvented by cutting the wall of the anther, although this does, of course, take a considerable amount of time.

Anthers can also be cultured in liquid medium, and pollen released from the anthers can be induced to form embryos, although the effi-ciency of plant regeneration is often very low. Immature pollen can also be extracted from developing anthers and cultured directly, although this is a very time-consuming process. Both methods have advantages and disadvantages. Some beneficial effects to the culture are observed when anthers are used as the explant material. There is, however, the danger that some of the embryos produced from anther culture will originate from the somatic anther tissue rather than the haploid microspore cells. If isolated pollen is used there is no danger of mixed embryo formation, but the efficiency is low and the process is time-consuming.

In microspore culture, the condition of the donor plant is of critical importance, as is the timing of isolation. Pretreatments, such as a cold treatment, are often found to increase the efficiency. These pretreatments can be applied before culture, or, in some species, after placing the anthers in culture. Plant species can be divided into two groups, depending on whether they require the addition of plant growth regulators to the medium for pollen/anther culture; those that do also often require organic supplements, e.g. amino acids. Many of the cereals require medium supplemented with plant growth regulators for pollen/anther culture. Regeneration from microspore explants can be obtained by direct embryogenesis, or via a callus stage and subsequent embryogenesis. Haploid tissue cultures can also be initiated from the female gametophyte.

In some cases, this is a more efficient method than using pollen or anthers. The ploidy of the plants obtained from haploid cultures may not be haploid. This can be a consequence of chromosome doubling during the culture period. Chromosome doubling may be an advantage, as in many cases haploid plants are not the desired outcome of regeneration from haploid tissues. Such plants are often referred to as 'di-haploids', because they contain two copies of the same haploid genome.

REGENERATION OF THE PLANT

In broad terms, two methods of plant regeneration are widely used in plant transformation studies, i.e. somatic embryogenesis and organogenesis.

Somatic Embryogenesis

In somatic or asexual embryogenesis, embryo-like structures, which can develop into whole plants in a way analogous to zygotic embryos, are formed from somatic tissues (Figure 2). These somatic embryos can be produced either directly or indirectly. In direct somatic embryogenesis, the embryo is formed directly from a cell or small group of cells without the production of an intervening callus.

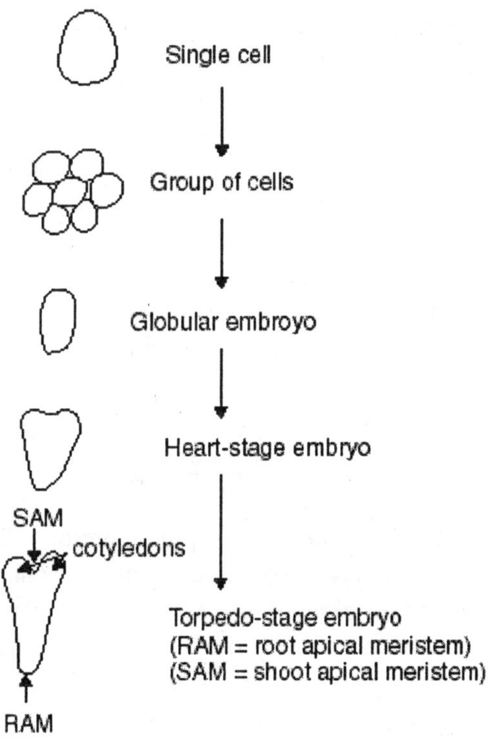

Figure 2: A schematic representation of the sequential stages of somatic embryo development.

Somatic embryos may develop from single cells or from a small group of cells. Repeated cell divisions lead to the production of a group of cells that develop into an organised structure known as a 'globular-stage embryo'. Further development results in heart- and torpedostage embryos, from which plants can be regenerated. Zygotic embryos undergo a fundamentally similar development through the globular, heart and torpedo stages. Polarity is established early in embryo development. Signs of tissue differentiation become apparent at the globular stage and apical meristems are apparent in heart-stage embryos.

Though common from some tissues, usually reproductive tissues such as the nucellus, styles or pollen, direct somatic embryogenesis is generally rare in comparison with indirect somatic embryogenesis. In indirect somatic embryogenesis, callus is first produced from the explant. Embryos can then be produced from the callus tissue or from a cell suspension produced from that callus. Somatic embryogenesis from carrot is the classical example of indirect somatic embryogenesis. Somatic embryogenesis usually proceeds in two distinct stages. In the initial stage (embryo initiation), a high concentration of 2,4-D is used. In the second stage (embryo production) embryos are produced in a medium with no or very low levels of 2,4-D. In many systems it has been found that somatic embryogenesis is improved by supplying a source of reduced nitrogen, such as specific amino acids or casein hydrolysate.

Organogenesis

Somatic embryogenesis relies on plant regeneration through a process analogous to zygotic embryo germination. Organogenesis relies on the production of organs, either directly from an explant or from a callus culture. There are three methods of plant regeneration via organogenesis. The first two methods depend on adventitious organs arising either from a callus culture or directly from an explant (Figure 3). Alternatively, axillary bud formation and growth can also be used to regenerate whole plants from some types of tissue culture.

Organogenesis relies on the inherent plasticity of plant tissues, and is regulated by altering the components of the medium. In particular, it is the auxin to cytokinin ratio of the medium that determines which developmental pathway the regenerating tissue will take. It is usual to induce shoot formation by increasing the cytokinin to auxin ratio of the culture medium. These shoots can then be rooted relatively simply.

An explant can be a variety of tissues, depending on the particular plant species being cultured. The explant can be used to initiate a variety of culture types, depending on the explant used. Regeneration by either organogenesis or somatic embryogenesis results in the production of whole plants. Different culture types and regeneration methods are amenable to different transformation protocols.

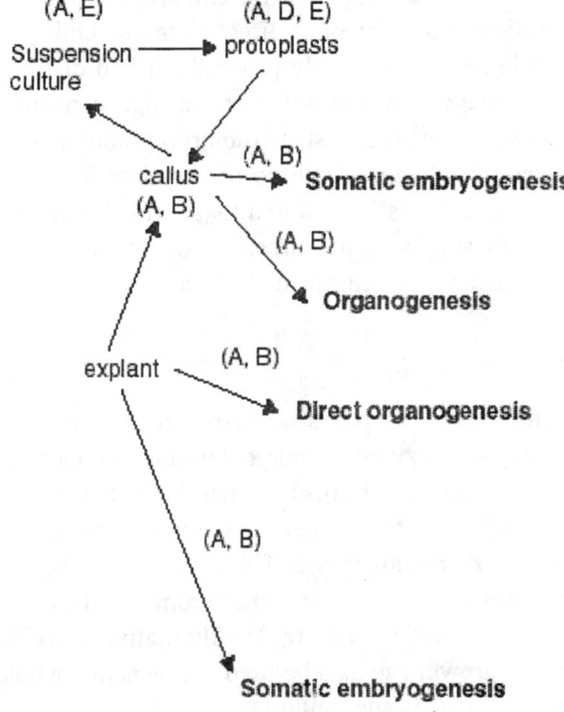

Figure 3: A simplified scheme for the integration of plant tissue culture into plant transformation protocols.

The transformation protocols highlighted in this figure are: (A) *Agrobacterium*-mediated; (B) biolistic transformation; (D) direct DNA uptake and (E) electroporation. Different combinations of culture type and transformation protocol are used depending on the plant species and cultivar being used. In some species a variety of culture types and regeneration methods can be used, which enables a wide variety of transformation protocols to be utilised. In other species there is effectively no choice over culture type and/or regeneration method, which can limit the transformation protocols that are applicable.

Chapter 5

Physical Components of Plant Tissue Culture Technology

A convenient location for a small laboratory can be a room or part of the basement of a house, a garage, a remodeled office or a room in the header house. The minimum area required for media preparation, transfer and primary growth shelves is about 14 m^2. Walls should be installed to partition different areas. Before setting up a commercial micropropagation unit, it is essential to check out the area keeping in mind the climate, and access to water, electricity, transportation, and infrastructure for supplies. A temperate climate is usually better suited to tissue culture ventures. This greatly reduces the cost of cooling required to maintain the temperature for optimum growth of the cultures.

The availability of electricity and water is of utmost importance, and should be taken into consideration while choosing the location of the facility. For example in India, of the 76 commercial tissue culture units, nearly 52 are located in and around the cities of Bangalore and Pune, where the climate is moderate. Hence, cooling is required only during certain periods of the year. However, in cities like Delhi, which have extremes of climate, tissue culture facilities require both heating and cooling. In a facility, which produces five million plants, the electricity cost per thousand plants is around US $0.30 in Bangalore and Pune; the same is about US $0.80 in Delhi.

Disruption of power and water supply causes major breakdown in the smooth running of tissue culture units. Poor quality water adds to the cost of media.

Location consideration should also include a check with local authorities about zoning and building permits before construction begins. The buildings should be located away from sources of contamination such as a gravel driveway, parking lot, soil mixing area, shipping dock, pesticide storage, and dust and chemicals from fields. Sanitation of the area is important in selecting location. The units should be located in areas where insect populations are minimal and air has low dust and pollen counts. In the case of export-oriented units, these should be near an international airport, which will reduce the time lag between packaging and shipment. This is critical to assure timely delivery of quality tissue culture products. For new ventures, the size of the facility should be kept small until the market acceptance is ensured.

Plant tissue culture laboratories have specific design requirements. Careful initial planning is, therefore, a prerequisite for successful running of a facility. The location and design of the laboratories should take into account isolation from foot traffic, control of contamination from adjacent rooms, thermostatically controlled heating and cooling, water supply and drains for a sink, adequate electrical service, provisions for a fan and intake blower for ventilation, and good lighting. Large sized facilities are frequently built free standing. Although more expensive to build, the added isolation from adjacent activities keeps the laboratory clean.

Prefabricated buildings make convenient low-cost laboratories. They are readily available in various sizes in many countries. Prefabricated buildings assembled on site can also be used. A single span building allows for a flexible arrangement of walls for dividing into convenient sized rooms. The floor should be of concrete or capable of carrying 170.5 kg/m2 (50 pounds per sq. foot). Walls and ceiling should be insulated to at least R-15, and covered inside with waterresistant material. Windows, if desired, may be placed wherever convenient in the media preparation and glassware washing rooms.

The heating system should be capable of maintaining room temperature at 200C during the coldest part of winter. A minimum of 2cm pipes should be used for water supply. Connection to a septic system or sanitary sewer should be provided. Air conditioning requirements should carefully estimated. Electrical service capacity for equipment, lights and future expansion should be calculated. For safety reasons, the electrical installation should be carried out professionally. Most electrical wiring will require 220 Volts, and autoclaves 230/250 Volts.

The areas such as the media preparation room, inoculation room and growth chambers should be isolated as 'clean zones'. The office, storage area, staff centre and packaging rooms can be maintained under ordinary conditions. The working areas must be demarcated according to the activities involved in the facility. Cleanliness is the major consideration when designing a plant tissue culture laboratory to minimise contamination. A positive pressure module should be installed to circumvent air intake from outside. Routine cleaning and aseptic procedures can decrease contamination losses to less than 1%. An enclosed entrance should precede the laboratories, and sticky mats should be placed to collect dirt from shoes.

The traffic pattern and workflow in the laboratory must be considered to maximise cleanliness. The cleanest rooms or areas are the culture room (aseptic transfer area) and the growth room. There should be no direct access to these rooms from outside. The media preparation area, glassware washing and storage areas should be located away from these rooms. The growth room and aseptic transfer rooms should be fitted with see-through doors and should be adjacent to each other.

Traffic through these areas should be minimal and restricted to the personnel working under laminar flow cabinets. Ideally, the media preparation area leads into the sterilisation area, which leads into the aseptic transfer room, and eventually to the growth room. Temperature and fire alarms must be connected directly to telephone lines to give fast warnings. An emergency generator should be available to operate essential equipment during power breakdown.

The basic equipment in most tissue culture facilities includes the following:

Autoclave: An autoclave is basically a large-sized but sophisticated pressure cooker, and is used for the sterilisation of the medium, glassware and instruments. Autoclaves of different sizes are available commercially. High-pressure heat is needed to sterilise media, water, and glassware. Certain spores from fungi and bacteria are killed only at 1210C and 1.05kg/ sq.cm (15 pounds per sq. inch) pressure. Self-generating steam autoclaves are more dependable and faster to operate.

Laminar airflow chamber: The laminar flow chambers provide clean filtered air that allows cultures to be handled under contamination-free environment. Several types of laminar flow chambers are sold on the market and are available in different sizes. The laminar-flow cabinets are located in the culture transfer area. Some large-sized laboratories have sterile rooms in addition to laminar flow cabinets.

Other equipment are; The sterilisation of instruments, such as the forceps, scalpel holders and blades is achieved with either gas flamed burners or with glass-bead sterilizers. The medium preparation room usually has the following equipment. A refrigerator-freezer to store chemicals and stock solutions, weighing scales for large amounts of over 10 g, and an analytical balance withl mg accuracy, a magnetic stirrer for the agitation, and a pH meter. Small laboratories may locate the refrigerator under the workbench to save space. High quality balances are essential in a tissue culture laboratory.

Most laboratories have top loading balances, which allow quick and efficient weighing. A hot plate with an automatic stirrer is needed to for preparing the media before autoclaving. The pH meter is needed to determine pH of the media. Some laboratories use pH indicator paper, however this method is considerably less accurate, and can severely affect the results. An aspirator can be attached to a water tap for filter sterilisation of chemicals and for surface sterilisation of the plant material. However, vacuum pumps are faster and more efficient, but also more expensive.

A drying oven is required to keep glassware such as beakers, flasks and cylinders, and is also useful for dry sterilisation of scalpels and glassware, such as Petri dishes, pipettes and others. The media containing carbon sources (e.g. sugars) and growth regulators are sterilised in the autoclave, but sometimes, aseptic filtration is better to avoid breakdown of heat-labile chemicals. The water still is also located in the medium preparation area. To prepare media, distilled or de-ionised water is generally used, although tap water can be used in some cases.

A variety of non-essential equipment is used in tissue culture laboratories. The specific requirements determine what need to be purchased. Microwave ovens are convenient for defrosting stock solutions and pre-heating agar media. Most laboratories have a dissecting microscope to excise small explants. Laboratory glassware washers or regular dishwashers can be used for replacing manual labour. Automatic media dispensers are helpful to pipette pre-set volume of media.

A gyratory shaker or a reciprocal shaker is necessary if micropropagation is based on liquid media or suspension cultures. Computers, photocopiers and fax machines are helpful for easy data management and maintenance of records. Some of the equipment may be costly, but goes a long way in saving time and labour and is essential for rapid communication in the competitive world.

Based on the different activities of a tissue culture, a facility can be divided into semiclean, clean and ultra-clean areas. The semi-clean areas comprise of the washing room, office and staff restrooms, where there is no need for maintaining sterile conditions. The clean areas encompass the media preparation and sterilisation rooms, which have to be sufficiently clean. High sterility has to be maintained in the culture transfer rooms and the growth rooms, which constitute the ultra-clean areas.

The glassware washing area should be located near the sterilisation and medium preparation rooms. This area should have at least one large sink but two sinks are preferable. Adequate workspace is required on each sides of the sink; this space is used for glassware soaking and drainage. Plastic netting can be placed on surfaces near the sink to reduce glassware breakage and enhance water drainage. The outlet pipe from the sink should be of PVC to resist damage from acids and alkalis. Both hot and cold water should be available and the water still and de-ionisation unit should be located nearby.

The choice of electrical washers should be based on the projected use, durability, reliability and cost, and service availability. In India and some other developing countries, where labour is relatively cheap, washing is done manually. The washing room should be swapped periodically. Mobile drying racks can be used and lined with cheesecloth to prevent water dripping and loss of small objects. Ovens or hot air-cabinets should be located close to the glassware washing and storage area. Dust-proof cabinets and storage containers should be installed to allow for easy access to glassware. When culture vessels are removed from the growth area, they are often autoclaved to kill contaminants and to soften semi-solid media. It should be possible to move the vessels easily to the washing area. The glassware storage area should be close to the wash area to expedite storage and access for media preparation.

The Store

It is advisable to have a separate area for storage of chemicals, apparatus and equipment. It would not only facilitate constant availability but also save cost from bulk purchase. Chemicals required in small amounts should not be purchased in large quantities as they may lose their activity, pick up moisture or get contaminated. Such problems can be overcome by purchasing small lots on a regular basis.

Media Preparation and Sterilisation

The media preparation room should have smooth walls and floors, which enable easy cleaning to maintain a high degree of cleanliness. Minimum number of doors and windows should be provided in this room but within the local fire safety regulations. This reduces cost and contamination. The media preparation and sterilisation can be carried out in the same area but preferably in different rooms, which need not be separated with doors. Media preparation area should be equipped with both tap and purified water. An appropriate system for water purification must be selected and fitted after careful consideration of the cost and quality.

A number of electrical appliances are required for media preparation; hence, it is essential to have safety devices like fire extinguisher, fire blanket and a first aid kit in the media preparation room. A variety of glassware, plastic ware and stainless steel apparatus is required for measuring, mixing, and media storage. These should be stored in the cabinets built under the worktables and taken out for use as and when required. This would save the cost and space for building storage shelves. The use of glassware should be kept at a minimum, as it will help in reducing losses due to breakage.

As far as possible, plastic ware and stainless steel vessels should be used, as they are much cheaper and more durable than glassware. The water source and glassware storage area should be in or near the medium preparation area. Work bench tops, suitable for comfortable working while standing should be 85 to 90 cm high and 60 cm deep. The workbench tops should be made with plastic laminate surfaces that can tolerate frequent cleaning.

Sterilising Room

The sterilising room should be in continuation with the media preparation room. The layout must be planned in such a way that it ensures the smooth movement of the containers from the washing to the media preparation and sterilisation room. The sterilisation room must have walls and floors that can withstand moisture, heat and steam. An exhaust should be fitted to remove the warm and moist air. The exhaust fan should have an outer cover to prevent entry of outside air. The fan cover should open only when the fan is in operation.

In small tissue culture facilities, costly autoclaves can be replaced by simple pressure cookers. However, for large volume media making, horizontal or vertical autoclaves should be installed. Double door autoclaves, which open directly into the media storage room, may be costly but reduce contamination. A cheaper alternative is to transfer the sterilised media to the adjoining room through a hatch window.

Transfer Room

The most important work area is the culture transfer room where the core activity takes place. The transfer area needs to be as clean as possible and be a separate room with minimal air disturbance. Walls and floors of the transfer room must be smooth to ensure frequent cleaning. The doors and windows should be minimal to prevent contamination, but within local safety code. There is no special lighting requirement in the transfer room. The illumination of the laminar airflow chamber is sufficient for work. Sterilisation of the instruments can be done with glass-bead sterilizers or flaming after dipping in alcohol, usually ethanol. The culture containers should be stacked on mobile carts (trolleys) to facilitate easy movement from the medium storage room to the transfer room, and finally to the growth room.

The chair seats of the transfer operators should be comfortable, as they have to work for long periods in the same place. Fire extinguishers and first aid kits should be provided in the transfer room as a safety measure. The personnel should leave shoes outside the room. Special laboratory shoes and coats should be worn in this area. Ultraviolet (UV) lights are sometimes installed in transfer areas to disinfect the room; these lights should be used only when people and plant material are not in the room. Safety switches can be installed to turn off the UV lights when regular room lights are turned on.

Growth Room

Growth room is an equally important area where plant cultures are maintained under controlled environmental conditions to achieve optimal growth. It is advisable to have more than one growth room to provide varied culture conditions since different plant species may have different requirements of light and temperature during in vitro culture. Also, in the event of the failure of cooling or lighting in one room, the plant cultures can be moved to another room to prevent loss of cultures. In the growth room, the number of doors should be minimal to prevent contamination. There is no need for windows in the growth room, except when natural light is used.

When artificial lighting is used, the external light can interfere with the photoperiod and temperature of the growth room. Depending on the amount of available space and cost, the culture containers can be placed on either fixed or mobile shelves. Mobile shelves have the advantage of providing access to cultures from both sides of the shelves. The height of the shelves should not exceed 2m. High shelving requires step-up stools to place and remove cultures, being dangerous and time consuming. The primary source of illumination in the growth room is normally from the lights mounted on the shelves.

Overhead light sources can be minimised, as they would be in use only while working during the dark cycle. Plant cultures may not receive uniform light from the conventional downward illumination. Lights directly fitted to the racks create uneven heat distribution. This leads to high humidity within the culture containers, which in turn can cause hyperhydricity. Sideways illumination is an alternative, which requires less number of lights, and provides more uniform lighting. But care has to be taken not to break the lights while moving the cultures across the shelves.

CONTROL OF GROWING CONDITION

Controlled temperature, lighting and relative humidity, and shelving need to be considered in planning the growth room. These vary depending on the size of the growth room, its location, and the type of plants cultured. For example, a small growth room located in the cool North American climate can be placed in an unheated or minimally heated basement. The chokes (ballasts) of the fluorescent lights need not to be separated; rather they can serve as a source of heat. Excess heat can be dissipated from the growth room, and used for heating other areas in the basement.

In such a situation, solid wooden shelves with space between shelves can be used and prevent culture vessels on shelf above the lights from becoming over-heated. However, a large growth room located above ground needs to have the light chokes installed outside the room. Shelves in large growth room can be of glass or metal wire mesh.

Temperature Control

Temperature is a primary concern in growth rooms; it affects decisions on installation of lights, control of relative humidity, and type of shelving. Temperature in the growth room is usually controlled with air conditioners. Generally, temperatures are kept around 220C. Heating is provided from conventional heating systems and can be supplemented with heat from light chokes. In most developing countries, cooling the growth room is usually a bigger problem than heating. Cooling can be provided with heat pumps, air conditioners and exhaust fans. Using open windows to cool culture rooms leads to contamination during summer and humidity problems in winter.

Lighting Control

Some plant cultures can be kept in complete darkness; however, most culture rooms need to be illuminated at 1 Klux [134.5 μmole/m²/s (microeinsteins per second per sq. centimetre or approximately 1076 foot candle) with some up to 5 to 10 Klux (672-1345 μmole/m²/s). The plant species and/or propagation scheduling determines the light intensity. The developmental stage of the plants also determines if wide spectrum or cool white-fluorescent lights are to be used. Rooting is strongly influenced positively with far-red light; therefore, wide spectrum lights should be used during Stage III and cool-white lights during Stage I and II.

Automatic timers are needed to maintain the desired photoperiod. Reflectors can be placed over bulbs to direct the light downwards and evenly. Heat generated by lights may cause condensation and temperature problems. Small fans placed at the end of the shelves increase airflow and decrease heat build-up. Reflective glossy paint on the walls provides an even light distribution as well as reduces the number of lights required. Relative humidity (RH) is difficult to control inside the culture vessels, but wide fluctuations in the growth room have a deleterious effect. Cultures can dry out if the room's RH is less than 50%. Humidifiers can be used to correct this problem. If the RH becomes too high, a dehumidifier is recommended.

Shelving

Shelves within the growth rooms vary depending upon the situation and the plants grown. Frames for the shelves can be made from 1.25cm (half-inch) thick angle iron. Shelves built from rigid wire mesh to allow maximum air movement and minimise shading should be used. Wood is inexpensive to build shelves. The wood for shelving should have smooth exterior, and should be painted white to reflect light. Expanded metal is more expensive than wood, but provides better air circulation. Tempered glass is sometimes used for shelves to increase light penetration, but it is more prone to breaking. Air spaces of 5 to 10cm between the lights and shelves decrease heat on upper shelves and reduce condensation in culture vessels. A room that is 2.4m high will accommodate 5 shelves, each 45cm. apart, when the bottom shelf is 10 cm above the floor.

Greenhouse Facility

A critical stage in plant tissue culture is the interim phase between the laboratory and field conditions. In vitro derived plants need to be gradually hardened to field conditions. Plant hardening is usually carried out under greenhouse that ensures high survival of the tissue-cultured plants in the field. There are three types of greenhouses: Ground to ground, Gable, and Quonset type. The most commonly used greenhouse is the Quonset type. It contains movable or fixed benches with hardening tunnels on them. The size of the greenhouse must be based on the scale of production. Greenhouse glazing can be of glass or fibreglass. Polyethylene films or sheets of polycarbonate or acrylic can also be used. Air inflated double polyethylene covering is the most economic.

Appropriate light, shading and blackout systems can be achieved with supplementary lighting. Drip irrigation systems, misting and fogging can be installed as needed. Greenhouses erected in warm climates should have fan-assisted drip pad cooling especially during summer. Greenhouses in colder climates need to be heated. Floor and bench systems can be used for heating and cooling the air. Low cost plastic pipes can be used to circulate warm air, which are adequate and cost effective.

Packaging Area

A separate area should be designated for packaging in a commercial tissue culture unit. Packaging materials such as cardboard cartons and labels should be stored in this area. The type of packaging of a particular plant depends greatly on the temperature zones through which the consignment has to pass from the point of shipment to its destination. For example, if the consignment of tissue-cultured plants is passing anywhere within temperate zones, it is only necessary to take care of the frost conditions and pack accordingly. But if the consignment is passing from a tropical country to a temperate country, it becomes necessary to take care of the temperature zones in different places through which the consignment passes.

Thus, the type of packaging of tissue-cultured products varies with plant and destination. For example, in a package of 22 kg, about 40,000 lily plantlets can be packed. However, in the same parcel, only 20,000 Spathiphyllum plantlets or 10,000 Syngonium plantlets can be packed because of the higher respiration rates in these species. The heat build-up is much more in Spathiphyllum and Syngonium plants than the dormant lilies. Similarly, only 8000 Gerbera or 7000 Cordyline plants can be packed in the same package.

Cooling material such as ice, dry ice and other material take up much of the packaging space. Before loading and shipping, the packed items should be properly counted and rechecked. The cartons containing the cultures for shipping to the customer should be properly labelled with the names and addresses of the consignor and the consignee, and the details of the commodity (storage temperature, handling, etc.). Adequate care must be taken while packaging large consignments, so that there is no disturbance or damage during transit. To prevent any sort of delay, ensure that the consignment is accompanied by documents such as invoice, packing list, import permit, phytosanitary certificate and Generalized System of Preferences (GSP).

Cost of a Tissue Culture Facility

The building of a tissue culture facility includes the cost of land, construction, electrical installation and plumbing. If the available funds are limited, well-designed laboratories can be established by modifying existing structures. For example, a two or three room house or a trailer with appropriate modifications can be converted into a medium sized micropropagation facility. Depending on the production and storage capacity, a facility can be small- less 100,000 plants, medium- 100,000 to 500,000 plants, or large-scale- 500,000 to 2,000,000 plants per annum.

Cost Reduction Strategies

For a commercial tissue culture unit to be successful, it is essential to constantly find means to increase the efficiency of production, and bring down the cost of production. A number of low-cost alternatives can be used to simplify various operations and reduce costs in a tissue culture facility.

Contamination control

The loss of cultures increases the cost of production. Contamination in cultures is caused from natural contaminants such as dust, air-borne particles, bacterial and fungal spores, fibres, and hair. Man-made contamination occurs mainly from body, clothing, and from faulty procedures in the laboratory. It has been estimated that each operator generates a minimum of

1 to 5 million particles (bigger than 0.5μM diameter) per minute. Modern, high quality clean rooms have environment control systems that minimise contamination. Wearing laboratory coats (which are mandatory in Class 100 clean rooms) reduces contamination from clothing, skin and hair.

Maintaining cleanliness in other working areas is equally important. There is continuous entry of contaminants with the worker and through air entering the growth room. A set routine procedure should be followed to reduce contamination in the laboratory; however, such procedures will vary with the location. In the absence of high cost systems, careful planning and appropriate design of the laboratory goes a long way in reducing contamination.

Cleaning the culture containers

Costly dishwashing machines for cleaning the culture containers should be replaced by manual washing if labour is relatively inexpensive. The washed culture containers can be dried in sun instead of costly hot air ovens. In small-scale laboratories, the autoclaves can be replaced by large sized pressure cookers, which are much cheaper. Instead of having one or two huge horizontal autoclaves, which generate hot air pockets in the sterilising rooms, it is better to have more of smaller vertical autoclaves, which keep the air cool. It is a normal practice to use costly aluminium foil for wrapping instruments for sterilisation. This can be replaced by stainless steel containers, which are autoclavable. Savings can also be made in media and containers.

Instead of Petri dishes as stage for manipulation of culture, stainless steel plates, ceramic tiles and brown wrapping paper can be used; all of these can be autoclaved. Ethanol for hand and workbench sterilization can be replaced by industrial spirit. The careless handling of inflammables used for sterilization can be hazardous. Glass bead sterilizers can be used to sterilize forceps and scalpels instead of the conventional flame sterilization using spirit lamps or gas cylinders. Commercial bleach has also been successfully used in several laboratories for bench and instrument sterilization to reduce cost, and to prevent fire hazard.

Culture maintenance

Plant cultures can be maintained in rooms with air conditioners and tube lights instead of highly priced plant growth chambers. The conventional method of downward illumination can be replaced by sidewise lighting systems, which not only reduces the number of lights but also provides more uniform illumination to the cultures. In the tropical and Mediterranean regions, the electrical lighting systems can be replaced by sunlight.

Most plants are seasonal in demand. To meet the requirements of extremely large number of plants, commercial production has to be backed by well defined working procedures and monitoring performance of the operation. Commercial laboratories should produce a range of plants for different seasons to maximise thc use of the facilities throughout the year. This lowers the unit cost of plant production. Large-scale micropropagation is a labour-intensive process, and therefore organising the availability of personnel is quite important. To reduce organisational problems in big units, the management structure must be well planned.

Supporting information systems such as inventory control, production scheduling, space utilisation and daily targets should be well defined. To produce quality products on large-scale, there should be good co-ordination among the technicians, supporting staff, supervisors and the researchers. The job description and reporting system should be very clearly stated. In addition, personnel selection and training is critical for successful large-scale production. The production needs to be periodically reviewed to meet needs of the customers.

The physical components required for the functioning of a commercial tissue culture laboratory can be either scaled up or down according to the interests of the propagator. The correct design of a laboratory, big or small, will help maintain asepsis, thereby increasing the efficiency of the unit and achieving a high standard of work. A micropropagation company in India converted a three-room apartment into a medium-sized tissue culture laboratory. Plant such as Spathiphyllum, Syngonium, Ficus, hosta, calla lily, gerbera, and cordyline, were produced on a commercial scale.

Delivery of the tissue-cultured plants to the tune of 2.5 million per year has been made from this unit by adopting various low cost alternatives. Similar tissue culture facilities, aptly called 'bio-factories', in villages of Cuba produce up to one million banana plants and 2.5 million sugarcane plants annually. Significant cost reduction for large-scale production can be achieved with automation and mechanisation. Although, automation of certain steps of micropropagation has been investigated for the past 20 years, its commercial use has not been adopted. The capital costs of such automated systems have prevented their application.

CHAPTER 6

PLANT CELL CULTURE PRODUCTIVITY

Several products were found to be accumulated in cultured cells at a higher level than those in native plants through optimization of cultural conditions. For example, ginsenosides by Panax ginseng, rosmarinic acid by Colleus blumei, shikonin by Lithospermum erythrorhizon, diosgenin by Dioscorea, ubiquinone-10 by Nicotiana tabacum were accumulated in much higher levels in cultured cells than in the intact plants.

However, many reports have described that yields of desired products were very low or sometimes not detectable in dedifferentiated cells such as callus tissues or suspension cultured cells. In order to obtain products in concentrations high enough for commercial manufacturing, therefore, many efforts have been made to stimulate or restore biosynthetic activities of cultured cells using various methods. The following are typical approaches that may increase productivity of cultured plant cells.

CULTURAL CONDITION OPTIMIZATION

A number of chemical and physical factors affecting cultivation have been tested extensively with various plant cells. These factors include media components, phytohormones, pH, temperature, aeration, agitation, light, etc. This is the most fundamental approach in plant cell culture technology. Since there are many reports and patents concerning optimization of cultural conditions in order to improve growth rates of cells and/ or higher yield of desirable products

Sucrose and glucose are the preferred carbon source for plant tissue cultures. The concentration of the carbon source affects cell growth and yield of secondary metabolites in many cases. The maximum yield of rosmarinic acid produced by cell suspension cultures of Salvia officinalis was 3.5 g/L when 5% of sucrose was used but it was 0.7 g/L in the medium containing 3% sucrose. Among a number of other components in the medium phytohormones such as auxins and kinetins have shown the most remarkable effects on growth and productivity of plant metabolites. In general, an increase of auxin levels, such as 2,4-D, in the medium stimulates dedifferentiation of the cells and consequently diminishes the level of secondary metabolites. This is why auxins are commonly added to the medium for callus induction, but they are added at a low concentration or omitted for production of metabolites.

Cytokinins stimulated alkaloid synthesis which was induced by removing auxin from the medium of a cell line of C. roseus. However, productions of L-DOPA by Mucuna pruriens, ubiquinone-10 by N. tabacum and diosgenin by Diocorea deltoidea were stimulated by high levels of 2,4-D. Although kinetin is one of the most popular cytokinins, 4-chloro-2-diphenylurea was reported to stimulate production of an antitumor compound, tripdiolide, by Triptorygium wilfordii cell cultures. Gibberellic acid is also effective on plant cell cultures. The growth of callus of a taxol-producing plant, Taxus cuspidata, was significantly promoted by addition of gibberellic acid into the solid medium.

Temperature, pH, Light and Oxygen: The effects of temperature, pH, light and oxygen are all parameters that must be examined in the studies secondary metabolites production. A temperature of 17- 25°C is normally used for induction of callus tissues and growth of cultured cells. But, each plant species may favor a different temperature. Lowering the cultivation temperature increased the total fatty acid content per cell in dry weight. The medium pH is usually adjusted to between 5 and 6 before autoclaving and extremes of pH are avoided. The optimum pH is determined and controlled using a small scale bioreactor or a jar fermentor with pH control equipment.

Present scale-up technology dictates the use of stainless steel tanks for growth of plant cells on an industrial scale, thus in general, eliminating the use of light. However, since there are cases of light-stimulated secondary metabolites production, this factor should be investigated as it could help elucidate regulatory factors. Modification of fermentors with lighting facilities have also been carried out. Various fermentors have been designed and tested by many researchers. These include modification of impellers or agitators for microbial fermentors, improvement of air-lift fermentors and use of new type reactors such as rotary-drum type fermentors. For hairy root cultures, fermentors equipped with special hangers inside the vessel are being used. Each plant species has different optimized conditions both for growth of the cells and for production of useful products, so it is necessary to optimize the conditions in each case.

High Cell Density Culture: To increase the productivity of secondary metabolites, high cell density cultures have been investigated. Using a newly designed fermentor and optimized culture medium, Coptis japonica cells were grown up to 75 g/ L of cell mass. The highest yield of berberine, 3.5 g/L, was produced intracellularly in 55 g/L of the cell mass.

Absorption of Products: Most products are generally accumulated intracellularly by cultured plant cells, but some compounds were reported to be secreted into the media. Chinchona ledgerina cells excrete anthraquinones in the liquid medium. Addition of a resin, XAD-7, into its suspension culture stimulated the production of anthraquinones up to 539 mg/L which was approximately a 15 times increase compared to the medium without resin. The pigments were mostly found to be absorbed by the resin. The yields of ajmalicine and serpentine produced by C. roseus were also increased by addition of XAD-7 and the ratio between both alkaloids produced was changed. It is of interest that production of these alkaloids which are known to accumulate inside cells were affected by the presence of resin. A similar approach is in which addition of active charcoal in the medium stimulated the yield of coniferyl alcohol up to 60-fold in a Maticaria chamomila culture. Increases of several plant products produced using a continuous extraction process with two-phase organic solvents.

High-Producing Strain Selection

The physiological characteristics of individual plant cells are not always uniform. For example, pigment producing cell aggregates typically consist of producing cells and non-producing cells. In 1976, some researchers in Germany obtained cell lines of Catharanthus roseus which accumulated higher levels of ajmalicine and serpentine as determined by radioimmunoassay. This is similar to monocolony isolation of bacteria. Following their excellent results, a number of researchers have used cell cloning methods as this is the most promising way of increasing the levels metabolites present. Some typical examples are shown in Table 1. Most of them are related to production of pigments, such as anthocyanins, as visual selection is easy because of the color.

Table 1: Typical examples of Cell Cloning Application

Products	Plants	Factors
Anthocyanins	Vitis hybrid	2.3-4
Anthocyanins	Euphorbia milli	7
Berberine	Coptis japonica	2
Biotin	Lavendula vera	9
Ubiquinone-10	Nicotiana tabacum	15

A strain of Vitis hybrid using an agar plating feeder method; the culture produced 3.4% of anthocyanins. A strain of Euphorbia milli was also recognized to accumulate about 7 times higher amounts of anthocyanins than that of the parent strain after 24 selections. Statistical and cell-pedigree analysis proved that production of the red pigments was stable. Investigation with C. roseus, anther, leaf, and meristem explants, synthesis and accumulation of alkaloids in cell cultures differ significantly provided the comparison is based on individual alkaloid types. Variation of alkaloid profiles in callus tissue developed from the anther walls and fillaments of C. roseus ranged from cell lines with no detectable alkaloids to those with 12 alkaloids representing Corynanthe-, Strychnos-, Aspidosperma-, and Iboga-type alkaloids.

Researchers repeated cell cloning using cell aggregates of Coptis japonica, and obtained a strain which grew faster and produced a higher amount of berberine and cultivated the strain in a 14 L bioreactor. The selected cell line increased growth about 6-fold in 3 weeks and the highest amount of the alkaloid produced was 1.2 g/L of the medium. The strain was very stable, producing a high level of berberine even after 27 generations.Cultured, green Lavendula vera cells grown in the light were found to accumulate a high level of free biotin. To select a high-producing cell line, pimelic acid, a precursor of biotin, was used as a selecting agent. The level of biotin accumulated by a selected cell-line was 0.9 μg/L which was 10 times the amount found in the leaves.

The selection of high pyrethrin producing tissue cultures derived from Chrysanthemum cineraliaefolium and indicated that analytical screening of the tissue lines enabled the selection of a few "high yielding" strains which were derived from high yielding plant selections. A rapid assay method is crucial in the selection of a high yielding cell line. A procedure using fluorescence assay to select ajmalicine and other major heteroyohimbine alkaloids producing cells of C. roseus was of clear advantage even over a radioimmunoassay procedure since almost unlimited numbers of colonies could be screened in this way. However, the cell cloning is especially compatible with selection for high pigment production since selection can be achieved visually, or with the usc of simple spectrophotometric analysis.

Cell cloning is undoubtedly a very useful technique to increase the level of secondary metabolites and it should be applied as widely as possible. However, it is not obvious why cultures contain both high- and low-yielding cells. Only a few papers concerned with possible mechanisms have appeared. The lack of specific enzymes represents the most important reaction for the inability of plant cell cultures to produce secondary metabolites. Protoplasts were also used for the selection of high-shikonin producing cell lines of L. erythroryzon and thiophene producing Tagetes patula cell lines.

Addition of Precursors and Biotransformation

Addition to the culture media of appropriate precursors or related compounds sometimes stimulates secondary metabolite production. This approach is advantageous if the precursors are inexpensive. Researchers initially examined the production of alkaloids with this approach in the 1960's, many similar experiments have been carried out. For example, amino acids have been added to cell suspension culture media for production of tropane alkaloids, indole alkaloids, and ephedorin and some stimulative effects have been observed. It is true that some amino acids are precursors of various alkaloids, but generally the biosynthetic steps from amino acids to alkaloids are so complicated that the author doubts whether amino acids added were incorporated into the alkaloids directly in cell culture.

Perhaps, they affected not only alkaloid biosynthesis directly as precursors, but also indirectly through other metabolic pathways in the cells. Phenylalanine is one of the biosynthetic precursors of rosmarinic acid. Addition of this amino acid to Salvia officialis suspension cultures stimulated the production of rosmarinic acid and shortened the production time as well. Addition of 500 mM tropic acid to the medium of Scopolia japonica increased the amount of alkaloids by up to 14 times. The level of an anticancer compound, tripdiolide, produced by T. wilfordii cultured cells was increased by addition of 100 μg/L farnesol which is dephosphorylated farnesyl pyrophosphate, and an intermediate in the biosynthesis of terpenoids.

Addition of phenylalanine into the agar medium of Taxus cupsidata cells was found to stimulate the biosynthesis of an anticancer compound, taxol. However, biotransformation using intact cells or immobilized cells is an alternative way of producing a product by adding precursors into the culture media. Rauwolfia serpentina formed a new indole alkaloid, 6-hydroxytaumacline, in significant amounts when the cells were cultivated in the presence of ajmalicine. Arbutin, a skin depigmentation agent, is produced by biotransformation of hydroquinone using C. roseus cells. Addition of the precursor into the liquid culture medium of this cell line produced arbutin efficiently.

In order to produce a prodrug, the glucosylation of umebelliferone and salicylic acid using Mallotus japonicus cells and reported that the latter was converted to its o-glucoside with a yield of 90-95%. The maximum level of the product obtained was 0.9 g/L. The glucoside showed as potent an analgesic activity as salicylic acid, while its effect was more rapid and more long-lived than that of salicylic acid in mice. It is of interest that plant cells are able to form prodrugs having commercial usefulness. The conversion of geraniol and nerol to neral and geranial by Vitis vinifera cell suspension cultures is also interesting. Using plant cell culture techniques, radioactive labelled compounds can be formed from appropriate substrates. This is a useful application of the technology because of high value of the product.

Instead of the addition of a particular compound as a precursor into the culture medium of plant cells, a suitable substrate compound may be biotransformed to a desired product using plant cells. This approach has been extensively applied in the fermentation industry using microorganisms and their enzymes. For example, L-aspartic acid and L-malic acid are being manufactured commercially from fumaric acid, respectively using microorganisms. And various steroids are also produced by microbial biotransformations.

Biotransformation of ß-methyldigitoxin to ß-methyldigoxin using D. lanata cells has been extensively investigated by some researchers in Germany since 1974 because digoxin has a large market as a cardiac glycoside. About 600-700 mg of ß-methyldigoxin per litre was obtained using a 200 L reactor. This process was studied for commercialization.

Processes for production of a very expensive antitumor drug, vinblastine, from catharanthine and vindoline using a biotransformation as well as a simple chemical synthesis. The process is now being developed for commercialization by Mitsui Petrochemical in Japan. Concerning vinblastine-related compounds production, and also reported that a biotransformation of catharanthine and vindoline to anhydrovinblastine using horseradish peroxidase and glucose oxidase mediated coupling of vindoline and catharanthine.

Biotransformation processes including addition of substrates into the cultures are one of the most commercially realistic approaches in plant tissue cultures because of economic reasons. However, the availability of inexpensive precursors is a key issue.

Microbial Elicitor Treatment

Microbial infections of intact plants often elicit the synthesis of specific secondary metabolites. The best understood systems are those of fungal pathogens in which case the regulatory molecules have been identified as glucan polymers, glycoproteins and low molecular weight organic acids. Examples of microbial elicitor induction include psoralen production in parsley diosgenin production in the Mexican yam and many others.

A cell line of Papaver somniferm that synthesized and accumulated sanguinarine, a quaternary benzophenanthridine alkaloid when exposed to a homogenate of the fungas Botrytis. A portion of the sanguinarine was released into the culture medium. Sanguinarine extracted from the intact plants is being used for oral hygienic products. Effects of elicitors on secondary metabolism have been investigated at the enzymatic levels to determine their mode of action.

A transient increase in rosmarinic acid, a-0-caffeoyl-3,4-dihydroxyphenyllactic acid, content in cultured cells of L. erythrorhizon after addition of yeast extract to the suspension cultures: a maximum was reached in 24 hr. When the plant cells were treated with yeast extract on the 6th day of the cultivation, the level of rosmarinic acid increased 2.5 times and the activity of phenylalanine ammonia-lyase in the cells rapidly increased before synthesis of rosmarinic acid.

Recent developments in phytochemical elicitation have shown that simple inorganic and organic molecules can induce product accumulation. Sodium orthovanadate and vanadyl sulphate induced the accumulation of isoflavone glucosides in Vigna angularis cultures and indole alkaloid accumulation in Catharanthus roseus cultures, respectively. Other substances found to stimulate alkaloid accumulation in C. roseus include sodium chloride, potassium chloride and sorbitol as well as abscisic acid. Processes such as these, employing simple and cheap elicitors have much promise in industrial scale plant cell cultures.

Addition of oxalate to the medium of Gossypium hirsutum suspension culture could reduce the amount of Verticillium dahliae elicitor to be employed to stimulate metabolite synthesis. Addition of a fungal elicitor often inhibits the growth of plant cells but a combination of the elicitor and oxalate did not reduce the cell mass of the plant, therefore secondary metabolite synthesis was increased up to ten fold. Researchers found that a combination of phosphate limitation and fungal elicitation synergistically increased production of secondary metabolites. They found that either phosphate limitation or elicitation with a mycelial extract of the fungus, Rhizoctonia solani alone results in increased production of the sesquiterpene solavetivone by Agrobacterium rhizogenes-transformed hairy root cultures of Hyoscyamus muticus. However, when phosphate limitation is coupled with fungal elicitation, the productivity increase is considerably greater than that obtained with either method alone.

Although the mechanism by which elicitors increase the productivity of secondary plant metabolites has not been elucidated, their stimulating activity is quite significant if an appropriate elicitor is chosen to stimulate synthesis of a particular product. However, the use of microbial elicitors may not be economical since an elicitor-producing microorganism should be cultivated in a fermentor separately from cultivation of plant cells using another fermentor. The fermentation cost for an elicitor-producing microorganism is not always inexpensive. In this sense, a simple and cheap compound should be employed as an elicitor.

Immobilized Cells Application: Immobilization of plant cells is considered to be of importance in research and development in plant cell cultures, because of the potential benefits that could be provided:

- The extended viability of cells in the stationary (and producing) stage, enabling maintenance of biomass over a prolonged time period;
- Simplified downstream processing (if products are secreted);
- The promotion of differentiation, linked with enhanced secondary metabolism;

— Higher cell density enabling a reduced bioreactor size, thereby reducing costs and the risk of contamination;
— Reduced shear sensitivity, especially with entrapped cells;
— Promotion of secondary metabolite secretion, in some cases;
— Flow-through reactors can be used enabling greater flow rates;
— Minimization of fluid viscosity increase, which in cell suspension causes mixing and aeration problems.

An immobilization system which could maintain viable cells over an extended period of time and release the bulk of the product into the extracellular medium in a stable form, could dramatically reduce the costs of phytochemicals production in plant cell culture. However, an immobilized system also has the problems described below:

— Immobilization is normally limited to cases where production is decoupled from cell growth;
— The initial biomass must be grown in suspension;
— Secretion of product into the extracellularly medium is imperative;
— Where secretion occurs there may be problems of extracellular degradation of the products;
— When gel entrapment is used, the gel matrix introduces an additional diffusion barrier.

Due to these problems, a system with commercial potential has not yet been developed in plant tissue cultures. However, various immobilization methods have been developed, ie., entrapment, adsorption and covalent coupling. Some preliminary results have been obtained with immobilized cells. Early work with C. roseus, showed that agar, agarose and carageenan were all suitable immobilization matrices suitable for maintenance of cell viability; but alginate was superior in terms of ajmalicine production. The accumulation of serpentine by C. roseus and anthraquinones by

Morinda citrifolia were both enhanced in the immobilized state when compared with freely suspended cells. It should be noted however, that the possibility that alginate acts as an elicitor of secondary metabolism cannot be ruled out. Agar has been shown to stimulate shikonin accumulation in L. erythrorhizon cultures.

Adsorption immobilization has been successfully used with a number of plant species. Capsicum frutescens cells immobilized on polyurethane foam produced 50 times as much capsaicin as suspension cells. Similarly, Solanum nigrum cells accumulated glycoalkaloids to levels exceeding those found in suspensions. Datura innoxia cells accumulated tropane alkaloids with a profile similar to that of the intact plant, whilst in free suspensions productivity was markedly suppressed. In general it appears that mild immobilization either through gel entrapment or surface adsorption enhances productivity and prolongs the viability of cultured cells.

Immobilized cells can also be used as biocatalysts for biotransformations. Such a system compares favourably with the use of freely suspended cells since, in the case of immobilization, the catalyst is theoretically reusable and the product is easily separated from the biomass. The most appropriate example is that of the 12-hydroxylation of ß-methyldigitoxin to ß-methyldigoxin with alginate-entrapped Digitalis lanata cells. The enzyme activity was maintained by the immobilized cultures for a period of 61 days. Furthermore, the product was located in the extracellular medium. Mild permeabilization of the cells may enable biotransformation rates to be increased. Polyurethane-immobilized C. frutescens cells fed capsaicin precursors produced this metabolite at levels of up to 10 times those of non-fed cultures.

Glass fibres can be used as a carrier of plant cells to produce useful plant metabolites. Papaver somniforum cells were immobilized on fabric of loosely woven polyester fibres arranged in a spiral configuration on stainless steel support frame to produce sanguinarine, an antibiotic in oral hygiene. The yield was 3.6 mg/g-fw. by immobilized cells and was more than twice as much as by suspension cells.

Product Secretion

Many plant products produced by cell cultures have been reported to be accumulated intracellularly. However, it may be possible to produce much higher level of product if it was secreted into the medium. This is because the product intracellularly accumulated sometimes inhibits its own synthesis by regulation mechanisms such as product inhibition and repression. For many immobilized plant cell systems to work it is essential that a significant amount of product is released into the medium. Two types of immobilization system with C. roseus i.e. gel entrapment in polysaccharide beads and in polyacrylamide sheets, have both exhibited alkaloid release by mechanisms that do not appear to be associated with losses in viability.

Capsicum frutescens cells immobilized on polyurethane released capsaicin entirely into the medium, although other species immobilized by the same method retained the product intracellularly. To enhance release, permeabilization of cell membranes has been attempted, but with only limited success. Brodelius tested five permeabilizing agents on three different species, and although product release was achieved, cell viability dropped in most cases. The exceptions were DMSO and Triton X-100, applied to C. roseus cells. Other attempts to permealize cells have also resulted in non-viable populations and the use of electroporation has also resulted in a viability decrease. The release of betanin by Beta vulgaris cells ultrasonicated for 20 to 60 seconds has been reported, with no apparent effect on cell viability.

Chenopodium rubrum cells, immobilized in alginate beads, secreted the red betacyanin pigment amaranthin into the medium. However, the pigment was subsequently degraded; chitosan and DMSO permitted further product release into the extracellular medium, but this was also accompanied by product degradation. Low concentrations of chitosan and DMSO incubated with the cultures for 96 hr did not appear to affect viability significantly, but a longer incubation period had a deterlerious effect.

The lack of an appropriate means of product release is a serious problem in terms of an industrial approach particularly based on immobilization of plant cells. It is perhaps necessary to examine in vivo physiological mechanisms of release. Thus, some researchers have been investigating the factors involved in vacuolar phytochemical storage. The accumulation of indole alkaloids in C. roseus vacuoles has been attributed to an ion-trap mechanism whereby the basic indole alkaloids are trapped in the acidic vacuole due to their positive charge at low pH, preventing diffusion across the tonoplast. This mechanism has also been demonstrated for quinoline alkaloid accumulation in Cinchona species. Active uptake mechanisms have also been reported for indole alkaloids in C. roseus vacuoles and isoquinoline alkaloids in Fumaria capreolata L.

The physiological significance of these two possible mechanisms is not yet clear, since there is obviously strong evidence for both. In terms of product release, it is pertinent to note that in cell cultures, an efflux of alkaloids was observed under certain conditions, indicating an equilibrium between the intracellular and extracellular compartments that could be perturbed by medium acidification with subsequent product release. The release of serpentine by C. roseus cells was observed when the cells were filtered and resuspended in fresh or conditioned medium and it was suggested that temporary membrane uncoupling was responsible.

Although most plant cell culture products were shown to be accumulated intracellularly, that almost all of taxol produced by Taxus brevifolia cell cultures was detected in the culture filtrate. This is an exceptional case in plant tissue culture research.

Mutagenesis

In the fermentation industry, induction of genetic mutant strains of microorganisms is ubiquitous, and auxotrophic and/or regulatory mutants are used extensively to produce a variety of products including amino acids, nucleotides, antibiotics, etc. However, mutagenesis has limited applicability to plant cell cultures, because of their diploid genetic make-up: the chance of obtaining a double mutation in a target gene is less than 1 in 106. Although, in principle, haploid plants can be produced from anther cultures; in practice haploid cell cultures tend to revert to the diploid state. This makes the chance of isolating over-producing cells from mutagen treatment of haploid cells very low. Furthermore, biosynthetic pathways of many secondary metabolites and their regulation mechanisms in higher plants are not always understood precisely, therefore, it is also difficult to know what kind of mutants should be induced in order to increase product synthesis.

However, π-fluorophenylalanine resistant cell lines of tobacco cell cultures and found that five lines of N. tabacum and five lines of N. glauca out of 31 resistant cell lines accumulated higher levels of phenolics. And also reported that a resistant strain of N. tabacum produced 6 to 10 times higher levels of cinnamoyl putrescine than that of the parent strain. Fig. 1 shows the comparison of enzyme activities involved in the biosynthesis of cinnamoyl putrescines in a parent strain, TX1 and a resistant strain, TX4. Enzyme activities in the π-fluorophenylalanine resistant strain were much higher than those of the parent. A parent strain of C. roseus produced catharanthine only in the production medium but its tryptophan analog resistant mutant accumulated the same alkaloid even in the growth medium. Among several types of analogues tested, 4-fluorotryptophan was found to be the most efficient of them.

Increase of metabolite levels using regulatory mutants is theoretically possible and selection of suitable analogues for this

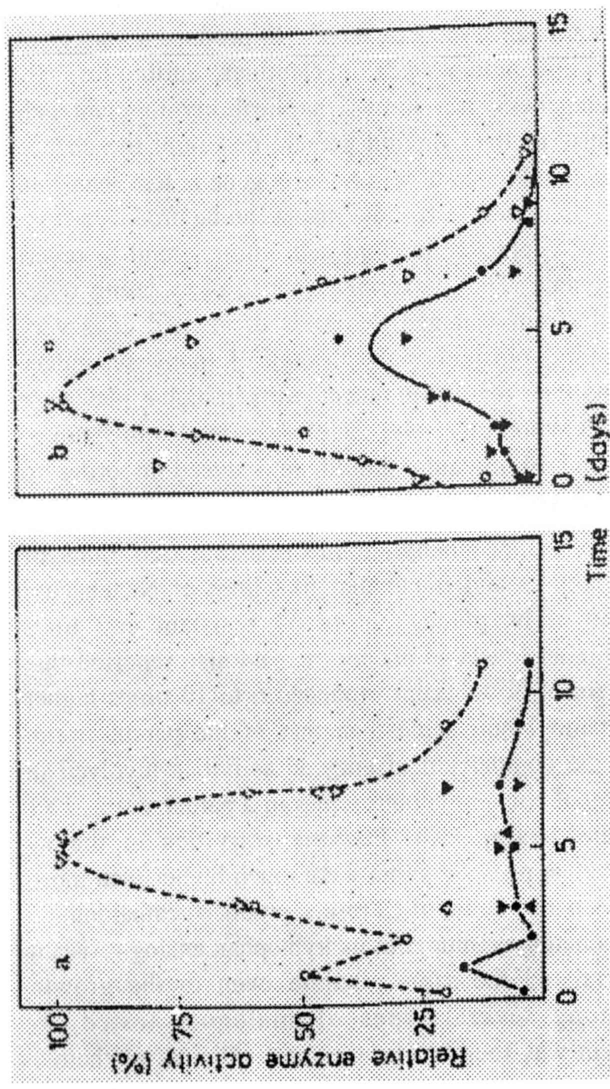

Figure 1: Comparison of Enzyme Activities involved in the Biosynthesis of Cinnamoyl Putrescines in TX1 (filled symbols) and TX4 (open symbols) of N. tabacum

purpose could be an important factor in order to produce a variety of products. Several research groups have used mutagens such as N-methyl-N'-nitro-N-nitrosoguanidine, ethylmethane sulfonate, N-ethyl-N-nitrosourea or X-rays to induce high yielding strains. In China, a fine callus strain of Anisodus acutangulus was derived through irradiation with 4000 R of X-ray, and the level of scopolamine in the cells was 0.177 mg/g d.wt., which was about 30% higher than of the parent. The productivity was stable.

A cell line of Lavendula vera producing a high level of free biotin was obtained by gamma ray irradiation for 1 hr. with 60Co. The line was found to contain 7 times the amount, 0.425 µg/g-fresh wt., of free biotin compared with the original unselected cells, 0.061 µg/g fresh wt., and 4.5 times that found in the leaves.In spite of the difficulty of obtaining auxotrophic mutants in plant cells, several groups have been actively working with this approach using haploid cells or protoplasts although their aim is not always for plant metabolite production.

Induction of a mutant having altered permeability could also be important, because plant cells generally accumulate their metabolites intracellularly, which is disadvantageous in commercial production because the amount of compounds produced is usually low. If the cells excreted products in large amounts, the product cost would be reduced. Thuja occidentalis excreted monoterpenoids, but the levels in the medium were only 5% of those in the mother plant. The same group reported that Macleya microcarpa cells excreted nearly all the alkaloids detectable in the culture flask. Some of the researchers cultivated Tinospora rumphii cells in 14 ml medium and found 0.57 (5.3% in dry weight cells) of isoquinoline alkaloids in the cells and 0.5 mg in the culture filtrate after 7 days cultivation. By replacement of the medium with fresh medium after 3 days of cultivation, 0.50 mg of the alkaloids in the cells, and 1.02 mg in the filtrate were accumulated. It seems then that the products must be excreted to some extent. Unfortunately, the secretion mechanisms for secondary products in higher plant cells have not yet been elucidated and extensive fundamental studies are required before meaningful manipulation can take place.

Organ Culture

It is desirable to use morphologically undifferentiated cells for the production of useful metabolites in ways similar to microorganisms and there are many examples which show high productive ability of such cells compared with intact plants. The development of a certain level of differentiation is considered to be important in the successful production of phytochemicals by cell cultures. There are many examples in the literature demonstrating a relationship between differentiation and secondary metabolic accumulation. Some researchers successively transferred the callus of Datura meteloides from a medium with auxins to one without and then cultivated continuously. As seen in Table 2, shoots, stems and roots were differentiated by turn and tropane alkaloid levels increased.

In Digitalis purpurea cultures, stimulation of digitalis cardenolides production by organ redifferentiation in callus tissues. A similar phenomena was also found in rotenone formation using Derris elliptica and morphinane alkaloid production using Papaver somniferum.

Table 2: Relationship between Productivity of Alkaloids and Differentiations

Plants	Alkaloid concentration (% dry wt.)
Callus	1×10^{-2}
Shoots-forming callus	1.5×10^{-2}
Growing shoots	2×10^{-2}
Roots-forming shoots	3×10^{-2}

Leaf of young plant	1×10^{-1}
Leaf of matured plant	1×10^{-1}

Researchers investigated the correlation between the stage of morphological differentiation and producing ability of the alkaloids using P. somniferumi, and found a green callus which differentiates epidermis or vascular bundles produced the alkaloids. A limited degree of tissue differentiation occurred and the cell contained codeine as a main alkaloid while the level of morphine increase as differentiation progressed.

Root cultures derived from suspension cultured cells of P. bracteatum were shown to produce thebaine in 0.03% yield. Axenic callus and shoot cultures of Pyrethrum cinerarifolium had an ability to produce pyrethrin. They isolated a few high yielding strains which were derived from high yielding plant selections. One isolate accumulated 11.3 mg total pyrethrins per 100 g dry weight but it subsequently differentiated into a shoot culture following the first analysis. Differentiated culture tended to produce more pyrethrins than did callus cultures. Using an established shoot culture derived from the disc floret of the plant, 341.8 mg of pyrethrins per 100 g wt., were obtained.

The cells were fractionated by Ficoll density gradient centrifugation. In the density fraction somatic embryos were formed in a medium containing 10-7 M zeatin but anthocyanin was scarcely produced. On the other hand, the cells in the lower density fraction synthesized anthocyanin in the same medium but formed few embryos. 40 to 50% of the total cells in the higher cell fraction synthesized anthocyanin at a maximum. The use of organs as opposed to cells or cell aggregates might necessitate an adaptation of cultured scale-up technologies, but in general these are not considered to be insurmountable. Indeed, root cultures of Panax ginseng have successfully been grown up to a volume of 20 KL and high levels of ginsenosides were obtained.

Hairy Root Culture

An alternative possibility is to induce biochemical differentiation, but suppress morphological differentiation. This would be difficult with an initially heterogenous population and the cells would have to be first selected for homogeneity in shape and metabolism. In fact, using a range of species, selection of regular green aggregates with a spherical shape and with cells of a regular morphology has been achieved. Such aggregates were found to enable high flow rates and easy medium removal, superior to those obtained with more dispersed cultures.

Two reports are worth noting with respect to differentiated cell growth and phytochemical production. It has been observed that the insertion of a very fine platinum or titanium wire into a disorganized callus of Helianthus tuberousus brings about morphological differentiation, whereas control, untreated tissue remained in an undifferentiated state. Should this phenomenon be reproducible in other species it might well serve as a well-defined means of controlling differentiation and possibly secondary metabolism. It was suggested that a charge transfer mechanism was responsible for these observations. A second discovery has shown that electric currents bring about shoot regeneration in callus culture at a rate five times greater than in controls. This has been patented as a process for the stimulation of growth and differentiation of plant tissue and for increasing the accumulation of secondary metabolites.

The use of Agrobacterium rhizogenes has been receiving attention recently in secondary metabolism research. It inserts the Ri plasmid into wounded tissue, causing the growth of very fine adventitious roots, so-called "hairy-roots". These roots can be cultured in hormone-free medium and there are several examples of enhanced accumulation of secondary products, relative to non-transformed tissue.

All hairy root clones of Hyoscyamus plants grew faster than ordinary root cultures and produced the similar level of tropane alkaloids to that accumulated in the intact plants. Some researchers shown that 17 day-old cultured hairy roots of Beta vulgaris had twice as much betacyanin and three times as much betaxanthin as seedling roots. On a mg/g dry wt. basis, the concentrations of these betalains were equal to or greater than those reported for storage roots. Another some researchers recently reported that hairy root cultures of Lupinus polyphyllus and L. hartweigii produced higher amounts of isoflavine glucosides than those of hormone-dependent cells. They also found that hairy root cultures of Peganum harmala synthesized higher levels of ß-carboline glucoid, ruine and serotonin.

Valeriana officinalis var. sambucifolia hairy roots produced 44.3 mg/g-cells d.w. of valepotriates having spasmolytic and sedative activities. The productivity was 0.9 mg/g/day and the concentration was approximately 4 times higher than that of normal roots. Researchers transformed Ruta gaveolens leaflets using A. rhizogenes and obtained the hairy root tissue. The tissue produced psoralen, isopimpirellin (methyoxylated furanocoumarins), xanthotoxin, bergapten, dictamunine, ?-fagarine, kokusaginine, edulinine and hydroxyrutacridone epoxide. The concentration of sucrose in the medium affected the levels of these secondary metabolites.

Hairy root culture systems for production of various chemicals. The levels of nicotine in Nicotiana rustica hairy roots and of betanine in Beta vulgaris were shown by his group to be at comparable or slightly higher than their intact plants. They also regenerated hairy roots from protoplasts of N. rustica hairy root tissues. Hairy root cultures of Datura stramonium and related Datura species to produce hyoscyamine as the major tropane alkaloid and small amounts of other tropanes including atropine and hyoscine (scopolamine). The biotechnological application of hairy root cultures is promising for a number of reasons: 1) stable, high level production; 2) fast auxin-independent growth; and, 3) the suitability for adaptation to fermentor systems. It is, therefore one of the most feasible techniques from an industrial point of view.

CHAPTER 7

CONCEPT OF TISSUE ENGINEERING

The application of the principles and methods of engineering and life sciences toward fundamental understanding of structure-function relationship in normal and pathological mammalian tissues and the development of biological substitutes for the repair or regeneration of tissue or organ function. Tissue Engineering differs from standard therapies in that engineered tissues become integrated within the patient, affording a potentially permanent and specific treatment of the disease state.

Tissue Engineering is an emerging multidisciplinary field that applies the principles of biology and engineering to the development of viable substitutes that restore, maintain or improve the function of human tissues. The different possibilities of creating replacement parts with TE were explored more than twenty years ago by Bell and colleagues, who investigated the possibility of the reconstruction of living tissues. During the 90s, TE progressed rapidly and biological substitutes were developed for several tissues in the body.

TE has emerged as a potential alternative to tissue or organ transplantation and tissue loss or organ failure may be treated either by implantation of an engineered biological substitute or alternatively with ex vivo perfusion systems. TE products may be fully functional at the time of treatment (e.g., liver assist devices, encapsulated pancreatic islets), or have potential to integrate and form the expected functional tissue upon implantation.

Currently the literature describes three general TE approaches. These principles are closely related to each other and may be applied to create new tissues. These approaches include:

— Design and grow human tissues in vitro for later implantation to repair or replace diseased tissues: The most common example is the skin graft, used for the treatment of burns. Skin graft replacements have been grown in tissue culture and used clinically for more than 10 years.

— Implantation of cell-containing or cell-free devices that induce the regeneration of functional human tissues: "signal" molecules, e.g. growth factors may be used to assist in biomaterial-guided tissue regeneration. Also, novel polymers have been created and assembled into three-dimensional configurations, to which cells attach and grow to reconstitute tissues. An example is the use of a polymer matrix to form cartilage.

— The development of external devices containing human tissues designed to replace the function of diseased internal tissues: This approach involves establishing primary cell-lines, placing the cells on or within structural matrices and implanting the new system inside the body. Examples of this approach include repair of bone, muscle, tendon and cartilage, endothelial cell-lined vascular grafts and heart valve substitutes.

The development of external devices containing human tissues (heart valves) designed to replace the function of diseased internal tissues.

TISSUE ENGINEERING OF HEART VALVES

TEHVs focuses on the development of a functional identical copy of a healthy heart valve. This is a relatively new technique within the TE-field and it may be possible to overcome all the disadvantages associated with the clinical use of mechanical and bio-prosthetic heart-valves. This technique may be able to improve the medical treatment of patients with heart-valve diseases.

A TE-approach for creating valve replacements may be categorized into two general strategies: (1) using degradable polymeric scaffolds or (2) acellular bio-matrices that support cell growth. The first strategy is to use degradable polymeric scaffolds moulded into heart valve geometries. Cells isolated from donor tissue are cultured and then seeded onto these scaffolds, resulting in constructs that can be implanted in vivo after a specific cultivation period.

The cells grow, develop, and produce extra cellular matrix (ECM) as the polymer degrades, ultimately leaving a natural tissue heart valve without any synthetic component. Ideally, autologous cells should be used to eliminate immunological responses to the TE-construct and to facilitate the growth and remodelling processes. The cell-polymer interaction is also critical because the quality and extent of ECM formation will determine the overall structure and mechanical properties of the newly developed tissue structure. The second TEHV strategy uses acellular, natural bio-matrices.

For example, porcine heart valves may be processed to remove their cellular antigens and reduce their immunogenicity. These constructs are then implanted in vivo and repopulated with host cells. This approach requires decellularization techniques that do not adversely affect the mechanical properties of the bio-matrices or the reconstitution of the tissue in vivo. Issues involving the stability and resorption of the natural bio-matrices must also be resolved. Supporters of this strategy argue, however, that in contrast to polymeric scaffold TEHVs, acellular biomatrices retain natural ligands and ECM constituents more suited for cell attachment and endothelialization.

Biomatrices: The second strategy, using decellularized biomatrices, also has great potential for fabricating TEHVs. The development of alternatives to the decellularization process in order to create scaffolds. Glutaraldehyde has been used since the 1960s to reduce the immuogenicity of xenogenic tissues. Using this agent, collagen fibers of the biomatrices are cross-linked to minimize xenogenic tissue solubility and antigenicity.

Also the mechanical properties of the natural biomatrix can be altered. Glutaraldehyde has been shown to increase the risk of heart valve calcification, and chemical residues from the fixation process can invoke an inflammatory response and reduce the viability of the repopulating cells. Due to this, alternative cross-linking solutions have been developed with different advantages and disadvantages. Decellularized porcine aortic valves and ovine pulmonary valves using a similar approach.

In both of these experiments native cells were removed from the biomatrices, which were subsequently repopulated with cultured cells. Additionally, the Steinhoff group implanted its TEHV constructs into lambs. Electron microscopy, histology and immunohistochemistry were used by both groups to evaluate the efficiency of their decellularization processes. The results suggested that the decellularization processes successfully removed most of the cells while preserving the ECM organization of the biomatrices.

The in vivo performance of TEHVs with echocardiography. Although the leaflets did thicken and calcify without an apparent loss of function, pulmonary regurgitation was not observed among the TE-constructs. A non-detergent-based decellularization solution and did not repopulate the biomatrices with cells prior to implantation into female sheep. Up to six months following implantation of these scaffolds, no pulmonary regurgitation, calcification or gross abnormality was observed. Host sheep cells repopulated the scaffold and the performance did not show significant difference from cryopreserved, cellularized, allogenic sheep aortic heart valves in comparable tests.

The decellularization method used by O'Brien et al., is the basis for the commercially available SynerGraft pulmonary replacement valve. CryoLife Inc, manufacturer of these valves, received a CE mark in October 2000 to distribute these heart valves throughout the European Union. Nearly six months later, the first successful implant of this valve was announced. A 3-year-old male child in Norway received the TE substitute. In comparison to polymeric scaffoldderived TEHVs, these replacements do not require pre-conditioning or pre-seeding.

Scaffold Materials

In the beginning of the 19th century, research into synthesized materials such as glycolic acid and other -hydroxy acids was abandoned, because the developed polymers were too unstable for long-term use. This instability, leading to biodegradation (BD), has proven to be immensely important in medical applications over the last three decades. Polymers prepared from glycolic acid and lactic acid have found a multitude of uses in medical practice. Since BD sutures were first approved in the 60s, diverse products based on lactic and glycolic acid and other related types have been accepted for use as medical devices.

In addition to these approved devices, a great deal of research continues on the biodegradability of these polymers. The design and development of TEHVs has benefited from many years of clinical use of a wide range of these polymers. Besides this, newly developed BD polymers and novel modifications of existing materials allow the creation of ideal scaffolds for many TE-applications.

In general, BD polymers can be categorized as biologically derived or synthetically produced. A BD polymer, which can be shaped into a heart-valve-scaffold and can be further cultured inside a BR after seeding. Both types of polymers under consideration for this purpose. Degradation and a processing technique called Fused Deposit Modelling that can process polymers into valvular scaffolds are reviewed as well.

BD Polymers: One of the current areas for applications of biodegradable polymers is TE. Several companies are investigating in the use of these materials as a scaffold to grow tissue on. Important properties in this regard include porosity for cell in-growth, a surface that balances hydrophilicity (affinity for water) and hydrophobicity (repelling in water) for cellular attachment, mechanical properties that are compatible with those of the tissue, and degradation rate and by-product production. To grow a TEHV, the polymer matrix may be utilized in three different ways:

— May represent the scaffold itself, which will degrade in vivo, where autologous cells grow over the structure.

— Can be a scaffold for cell growth in vitro that is degraded by the growing cells before the structure is implanted.

— Can be a combination of the first two.

In addition to these three possibilities, the scaffold can also be formulated to contain additives or active agents for more rapid tissue growth. In the future, device designers, tissue engineers and physicians will even have a wider choice of BD polymers as scaffolds for TEHV-applications, as biodegradable polymers are under constant development.

Biodegradable polymers are either derived from natural or synthetic sources. In general, synthetic polymers offer greater advantages compared to natural materials. Synthetic polymers can be tailored to give a wider range of properties and more predictable lot-tolot uniformity than can materials from natural sources. Synthetic polymers also represent a more uniform source of raw materials and are free from concerns of immunogenicity. Naturally occurring hydroxy acids, such as glycolic, lactic and å-caproic acids have been utilized to synthesize an array of useful BD polymers for variety of medical product applications.

The selection of the scaffold material plays a key role in the design and development of a particular TE product. Although the classical selection criteria focus on a safe, stable implant, it is recognized that every material used will elict a different cellular response in terms of degradation. One of the current challenges in culturing TEHVs is to select a polymer scaffold that meets the mechanical properties and degradation times required. The ideal polymer for a particular application should be configured so that it possesses the following properties:

— Appropriate mechanical strength to mimic in vivo conditions.

— Rate of matrix regeneration close to biodegradability rate of the BD polymer scaffold.

— Does not invoke an inflammatory or toxic response.

— Is metabolised in the body after fulfilling its purpose, leaving no trace (bioabsorbable).

— Is easily processable into the final product form, either porous or compatible with a range of extremely hydrophilic additives (starch, salt) to create porosity.

— Demonstrates acceptable shelf life and is easily sterilized.

In general, the factors affecting the mechanical performance of biodegradable polymers include monomer selection, initiator selection, process conditions, and the presence of additives. These factors in turn influence the specific features of the polymer such as: hydrophilicity, crystallinity, melt and glass-transition temperatures, molecular weight, molecular-weight distribution, end groups and presence of residual monomer or additives.

In addition, these properties of BD materials should be evaluated, in order to determine its effect on biodegradation. In general, an unstable backbone of the material leads to biodegradation. The most common chemical functional groups with hydrolytically unstable linkages are esters, anhydrides, orthoesters and amides. A review of natural and synthetic polymers that may be of use as TEHV-scaffolds.

Natural BD polymers

There arc many different existing potential biodegradable-scaffold materials that may be used for TE-applications. Five of the most commonly used materials that may be used as a tissue scaffold. These are:

Type I collagen: Collagen is the major protein component of mammalian connective tissue, accounting for 30% of all protein in the human body. It is found in every major tissue that requires strength and flexibility. Nineteen types of collagens have been identified; the most abundant being type I, which makes up more than 90% of all fibrous proteins. Individual collagen molecules, consisting of a chain of amino acids, that polymerise in vitro into strong fibers.

These fibers consist of three chains of amino acids that can be subsequently formed into larger organized structures like scaffolds. Cross-linking or chemical bonding can be enhanced after isolation through a number of well-described physical or chemical techniques. Increasing the intermolecular cross link's a) increases biodegradation time, b) increases hydrophobicity, c) decreases the solubility, d) and will increase the tensile strength of the collagen fibers.

Glycosaminoglycans (GAGs): GAGs are highly negatively charged molecules, with an extended conformation that gives high viscosity. GAGs are located primarily on the surface of cells or in the extra cellular matrix (ECM). Structural components of the ECM, such as collagen and GAGs, have major roles in valvular degeneration and calcification of bioprosthetic heart valves. In particular the GAGs located in the spongiosa layer of the heart valves are extremely important for mechanical properties.

Their rigidity provides structural integrity to cells and provides channels between cells that permit cell migration. The specific GAGs of physiological significance are hyaluronic acid, dermatan sulphate, chondroitin sulphate, heparin, heparan sulphate, and keratan sulphate. Due to its relative ease of isolation and modification and its ability to form solid structures, hyaluronic acid has become the principled GAG investigated for medical device development.

Chitosan: This biosynthetic polysaccharide can be slowly depolarised in vivo with lysozyme. Lysozyme is an enzyme that occurrs naturally in egg white, human tears, saliva, and other body fluids. It is capable of destroying the cell walls of some bacteria and acts as a mild antiseptic. The biodegradation time is determined by the amount of residual acetyl content, a parameter that can be easily varied. Chemical modification of chitosan produces material with a variety of physical and mechanical properties. Like hyaluronic acid, chitosan is non-antigenic and is a well-tolerated implant material. It can be formed into membranes and matrices suitable for several TE-applications.

Polyhydroxyalkanoates (PHA): PHA polyesters are degradable, biocompatible, thermoplastic materials produced by several different microorganisms. Depending on growth conditions, bacterial strain, and carbon source the molecular weight of these polyesters can range from tens into hundreds of thousands. The most extensively studied PHA is the simplest: Poly-3-hydroxybutyrate (P3HB). Most of these are homopolymers and are highly crystalline, extremely brittle and relatively hydrophobic. Consequently, these polymers can have in vivo degradation times in the order of years and therefore not suitable as a scaffold material.

P3HB and its copolymers containing = 30% 3-hydroxyvaleric acids are currently commercially available. It has been reported that a PHA copolymer of 3-hydroxybutyrate combined with 10% 3-hydroxyvalerate may provide an optimum balance of strength and toughness for a wide range of scaffold applications. It has low toxicity, partly due to the fact that it degrades in vivo to d-3-hydroxybutyric acid, a normal constituent of blood. Poly-4-hydroxybutyrate (P4HB) contains many of the same properties as P3HB and is an easily mouldable thermoplastic that can be formed into functional valve scaffolds.

Synthetic BD polymers

The synthetic biodegradable polymers that are currently used or being investigated for use in wound closure, orthopaedic fixation devices, dental applications, intestinal applications and cardiovascular applications. Most of the commercially available biodegradable devices consist of polyesters composed of homopolymers or copolymers of glycolide or lactide. The five most commonly investigated synthetic polymers used as matrices for TEHVs are described.

— *Poly (glycolic acid, poly (lactic acid) and their copolymers*: PGA, PLA and their copolymers are the most widely used BD polymers in medicine. Of this family of linear aliphatic polyesters, PGA has the simplest structure. This group is derived from organic compounds where carbon and hydrogen molecules are arranged in straight or branched chains, a type of hydrocarbon that includes; alkanes, alkenes, and alkynes.

— *Polyglycolide (PGA)*: PGA was used to develop the first totally synthetic absorbable suture, marketed as Dexon in the 1960s. Glycolide monomers are synthesized from the dimerization of glycolic acid. PGA is highly crystalline, with a high melting point and a glass-transition temperature between 35 and 40°C.

The glass-transition temperature is the temperature where a plastic changes from being brittle and hard into flexible material without changing phase. Because of its high degree of crystallization, it is not soluble in most organic solvents. Fibers from PGA exhibit high strength and modulus and are too stiff to be used as sutures except as a braided material. Due to its hydrophobic nature, scaffolding materials made of PGA tend to lose their mechanical strength rapidly, over a period of 2 - 4 weeks post implantation and are completely absorbed in 4 - 6 months.

— *Polylactide (PLA)*: PLA is more hydrophobic because of the extra methyl group in lactic acid and is therefore more soluble in organic solvents than PGA. This polymer exists as two optical isomers, d and l. l-lactide is a naturally occurring isomer. Dl-lactide is the synthetic blend of d-lactide and l-lactide.

The homopolymer of l-lactide (LPLA) is a semi crystalline polymer. These types of materials exhibit high tensile strength and low elongation. Consequently, these materials have a high modulus which makes them more suitable for load-bearing applications such as the cyclic stress experienced in the cardiovascular system. Poly dl-lactide (DLPLA) is an amorphous polymer exhibiting a random distribution of both isomeric forms of lactic acid, and accordingly is unable to arrange into an organized crystalline structure.

This material has lower tensile strength, higher elongation, and a much more rapid degradation time, making it attractive as a drug delivery system. The degradation time of LPLA is much slower than that of DLPLA, requiring more than two years to be completely absorbed. Copolymers of l-lactide and dl-lactide have been prepared to disrupt the crystallinity of l-lactide and can accelerate the degradation process.

— *Poly (lactide-co-glycolide)*: Using the polyglycolide and poly (l-lactide) properties as a starting point, it is possible to copolymerize the two monomers to extend the range of homopolymer properties. Copolymers of glycolide with both l-lactide and dl-lactide have been developed for both device and drug delivery applications. It also may be used as a scaffold material. It is important to note that there is not a linear relationship between the copolymer composition and the mechanical and degradation properties of the materials.

For example, a copolymer of 50% glycolide and 50% dl-lactide degrades faster than either homopolymer.

Poly -caprolactone: Poly -caprolactone (PCL) is an aliphatic polyester that has been intensively investigated as a biomaterial. The discovery that PCL can be degraded by micro-organisms led to the evaluation of PCL as a biodegradable packaging material. The ring-opening polymerisation of -caprolactone yields a semi crystalline polymer with a melting point of around 60°C. The polymer is regarded as tissue compatible and used as biodegradable sutures in Europe. Because the homopolymer has a degradation time in the order of two years, copolymers have been synthesized to accelerate the rate of bio-absorption. Copolymers of -caprolactone with dl-lactide have yielded materials with more-rapid degradation rates.

Poly-dioxanone: The ring-opening polymerisation of p-dioxanone resulted in the first clinically tested monofilament synthetic scaffolding material, known as PDO. This material has approximately 55% crystallinity, with a glass-transition temperature of -10 - 0°C. The polymer is processed at the lowest possible temperature to prevent depolymerisation back into a monomer. Poly (dioxanone) has demonstrated no acute or toxic effects after implantation. The monofilament loses 50% of its initial breaking strength after three weeks and is absorbed within six months.

Polyglyconate: Copolymers of glycolide with trimethylene carbonate (TMC), called polyglyconate, have been used as sutures, tacks and screws. These materials have better flexibility than pure PGA and are absorbed in approximately seven months. Glycolide has also been polymerised with TMC and p-dioxanone to form a suture that absorbs within 3 - 4 months and offers reduced stiffness compared to pure PGA fibers.

Degradation

Once implanted, a scaffold material should maintain its

mechanical properties until it is no longer needed and then be absorbed and excreted by the body, leaving no trace. A simple chemical hydrolysis of the hydrolytically unstable backbone is the prevailing mechanism for a polymer's degradation. This occurs in two phases. In the first phase, water penetrates the bulk of the device, attacking the chemical bonds and converting long polymer chains into shorter water-soluble fragments.

This occurs in the amorphous phase and initially there is a reduction in molecular weight without a loss in physical properties, since the device matrix is still held together by the crystalline regions. The reduction in molecular weight is followed by a reduction in physical properties, as water begins to fragment the device. In the second phase, enzymatic attack and metabolization of the fragments occurs, resulting in a rapid loss of polymer mass. This type of degradation, where the rate at which water penetrates the device exceeds that at which the polymer is converted into water-soluble materials, is called bulk erosion. This results in erosion throughout the device. All commercially available synthetic devices and sutures degrade by bulk erosion.

A second type of biodegradation, known as surface erosion, occurs when the rate at which the water penetrates the scaffold is slower than the rate of conversion of the polymer into water-soluble materials. Surface erosion results in the device thinning over time while maintaining its bulk integrity. In general, this process is referred to as bioerosion rather than biodegradation. The degradation-absorption mechanism is the result of many interrelated factors that include:

— Chemical stability of the polymer backbone.

— Presence of catalysts, additives, impurities, or plasticizers.

— The geometry of the device.

Factors that accelerate polymer degradation include an increased hydrophilic backbone, increased number of reactive hydrolytic groups in the backbone, less crystallinity, increased porosity and a larger surface area. Balancing each of these factors will allow an implant to slowly degrade and transfer stress at an appropriate rate to surrounding tissues as they heal.

Comparison of scaffolding materials

Biodegradable polymers that have been used in attempts to TE heart valves in vitro. Shinoka et al., attempted to identify a suitable degradable polymer for a fully functional and autologous TEHV. This group successfully constructed a TEHV-leaflet from woven and non-woven PGA mesh sheets. Leaflets from polyglactin sheets sandwiched between PGA mesh sheets were also fabricated. These materials were too thick, non-pliable and therefore could not form non-stenotic, tri-leaflet heart valves.

Furthermore, the fibrous PGA meshes had insufficient strength to withstand in vivo flow conditions. As a result, several naturally occurring thermoplastic polymers, like PHA and P4HB, where investigated for TEHV scaffold fabrication. All naturally thermoplastic polymers posses biocompatible, resorbable and flexible features. Furthermore, they have high mechanical strength and induce a minimal inflammatory response. The low melting point of these thermoplastics permits moulding into the configuration of a trileaflet heart valve, and salt leaching technique can be used to construct a porous scaffold that promotes cell ingrowth.

These polymers have been used either alone or in combination to fabricate different TEHV scaffolds. In a comparative study conducted by Sodian et al., the in vitro performance of TEHVs constructed from tri-leaflet-shaped polymer scaffolds of PHA and P4HB were examined. No significant difference was demonstrated. Increased cellularity and collagen formation was observed when compared to fibrous PGA sheets. Therefore, polymer scaffolds were developed to combine the favourable cell-polymer interactions of PGA with the processability and strength of thermoplastics.

Used scaffolds where PGA was moulded around a softened PHA tube in order to form the conduit wall. After this procedure leaflets constructed from PGA-PHA-PGA sandwiches were attached. Hoerstrup et al., presented another approach. Here the TEHV-scaffold was created from coated PGA meshes with a thin layer of P4HB. After the solvent evaporated, the P4HB-coating physically bonded with the PGA fibers. Finally, attempts were made to construct TEHVs from a nonporous PHO/PGA.

Conduit wall is fabricated from an outer layer of PGA pressed around an inner layer of PHA. Leaflets are fabricated from PGA-PHA-PGA sandwiches. Stock et al., fabricated valve conduit walls from nonporous PHO films sandwiched between nonwoven layers of PGA. In Figure 3.6a the superior view of this construct is shown. The leaflets were constructed from porous PHO and sutured to the conduit wall with polydioxanone (PDO). This construct consisted of four different biomaterials with different biomechanical, biochemical, and degradative properties.

The nonporous PHO wall inhibited cell ingrowth, and scar formation was observed on the exterior surface of the TEHV construct. Based on these observations, a new thermal processing technique was developed to replace leaflet suturing and to construct both the conduit wall and leaflets from porous PHO. The resultant porosity permitted cells to grow into the polymer, formed viable tissue, and initiated polymer degradation.

Scaffold processing

Different processing techniques have been developed for the design and fabrication of three-dimensional (3D) scaffolds suitable for TE implants. Conventional techniques for scaffold fabrication include fiberbonding, solvent casting, particle leaching, membrane lamination and melt molding. One of the newest methods being developed by Therics (Princeton, NJ) uses a system for building three-dimensional devices for use as scaffolds and for drug delivery products. In this system, small spheres of polymer are deposited as thin films.

Using technology similar to that found in ink-jet printers, small amounts of solvent are used to fuse particles together. The particles not fused are removed and another layer of particles is deposited. This particle placement and fusing is continued for many layers, until the exact three-dimensional structure is obtained. Because each polymer layer is applied in a separate step, different polymers can be used to obtain different properties of the interior and exterior surface of the device. Swinburne University has a fused deposition-modelling machine that may be used to fabricate porous scaffolds from BD materials.

Fused Deposition Modelling (FDM): FDM is a rapid prototyping process that integrates Computer Aided Design (CAD), polymer science, computer numerical control, and extrusion technologies to produce 3-D solid objects directly from a CAD model using a layer-by-layer deposition of molten thermoplastics extruded through a very small nozzle. This technique has been used to fabricate 3D scaffolds with a honeycomb structure from PCL. These scaffolds have good mechanical properties and a porosity around 60%.

FDM is one of the few commercially available rapid prototyping technologies that offers the possibility of producing solid or porous objects in a range of different materials including metals and composites. The FDM system, developed by Stratasys Inc, fabricates structures from different kind of plastics and BD polymers. One FDM machine, the FDM3000 can be used to fabricate scaffolds for TE applications with layer thicknesses ranging from 0.178±0.127 – 0.356±0.127 mm. The FDM method involves the melt extrusion of filament materials through a heated nozzle and deposition as thin solid layers on a platform.

The process begins with the creation of a solid model or a closed surface model with CAD. The model is converted into an STL file using a specific translator on the CAD system. The

STL file is then sent to the FDM slicing and pre-processing software called QuickSlice, where the designer selects proper orientation, creating supports and slicing and other parameters to prepare the part program for sending to FDM machine. A proper orientation of STL model is necessary to minimise or eliminate supports. The STL file is then sliced into thin cross sections at a desired resolution, creating an SLC file.

Each slice must be a closed curve and any unclosed curves are edited and closed. Supports are then created if required, and sliced. Supports can also be created as part of the CAD model and imported as part of the STL file. The sliced model and supports are then converted into SML file, which contains actual instructions code for the FDM machine tip to follow specific tool paths, called roads, to deposit the extruded material to create each cross section. The designer selects various sets and road parameters to create a SML file. The SML file is sent to the FDM machine and the FDM head creates each horizontal layer by depositing molten extrudea material on a foam foundation until the part is completed. The part is then removed, supports are detached carefully, and it is ready for use.

Cells Seeding of Scaffolds

Heart valves are composed of many different types of cells that include endothelial cells, myocytes and fibroblasts. To create a BD-scaffold that is a functional, durable TEconstruct, the establishment of cultured cells is a priority. To avoid rejection, cells should be autologous. The use of autologous cells has many advantages, including ethical considerations. Another cell source may be stem cells, but currently ethical issues make it impractical to use these cells for research purposes. In addition to this, to the authors understanding, the present lack of knowledge on how to trigger the mechanism to grow a particular organ or cell makes the use of this source an unrealistic possibility.

However, the stem-cell approach shows great potential for future TE-applications. A brief overview is presented of the basics of stem cell engineering. Futhermore, an overview is presented of the different available cell sources and how different cells react in vitro to simulated in vivo conditions. During this experimental investigation attempts were made to establish primary cell lines.

Cell Sources

Cells may be isolated from several sources but to grow a TEHV that can be directly implanted into a patient without rejection, investigators are limited to a few possibilities:

— Cells donated by other individuals (allogenic)

— Cells obtained from the same individual (autologous)

— Universal stem cells

Allogenic cells: Allogenic cells are cells that are isolated from a donor of the same species. Animal cells have been widely used for experimental cardiovascular implants and the use of human cells for in vitro investigations possesses ethical constraint. Allogenic dermal fibroblasts have been shown to be acceptable immunologically, and a source is to some extent available, from human foreskins. Dermal fibroblasts have proven inferior to vascular fibroblasts for vascular prostheses. Besides fibroblasts, endothelial cells from the human umbilical vein may be a suitable source for TE arteries.

Autologous cells: Therapies that use a patient's own cells are safest from an immunologic point of view. However, these methods may not always be available. For example, many surgeons are not enthusiastic about performing two operations, (i.e. one to harvest cells, and another, weeks later, to implant a cell seeded scaffold) because of the additional costs and time issues. Even when harvesting a patient's cells for immediate implantation there are two surgical sites, i.e. the implantation site and the harvest site. In these cases, there may be donor site morbidity, including infection and chronic pain, as well as additional surgical costs.

Finally, a very ill or elderly patient may not have sufficient viable cells, to establish useful cell lines. For each of these reasons, there is significant interest in having an "off-the-shelf" supply of donor cells. These cells would be expanded in vitro and immortalized. Foetal or neonatal tissues are extremely useful for this purpose since they are non-immunogenic and are a rich source of stem cells. This approach, however, is a very controversial ethical issue.

Stem cells: Stem cells have the ability to divide in culture and give rise to specialized cells. In order to understand the potential of stem cells a short overview is presented. A human fertilized egg is a totipotent cell, meaning that its potential is total. In the first hours after fertilization, this cell divides into two identical totipotent cells. Approximately four days after fertilization, these totipotent cells begin to specialize, forming a blastocyst, a hollow sphere of cells. The outer layer of cells will go on to form the placenta and other supporting tissues needed for foetal development in the uterus.

The inner layer of cells in the blastocyst will form virtually all of the tissues of the human body. These cells are pluripotent and give rise to many types of cells and tissues. The pluripotent stem cells undergo further specialization into stem cells that are committed to give rise to cells that have a particular function. Examples of this include blood stem cells that give rise to red blood cells, white blood cells, platelets and skin stem cells that give rise to the various types of skin cells.

These more specialized stem cells are called multipotent. While stem cells are extraordinarily important in early human development, multipotent stem cells are also found in children and adults. For example, one of the best-understood stem cells is the blood stem cell. Blood stem cells reside in the bone marrow of humans and perform the critical role of continually replenishing the supply of blood cells throughout life. Populating a TE scaffold with adult human stem cells may be possible.

Cell types

A number of different cell sources have been used to seed various polymer constructs for TEHV fabrication. Most of the experiments used mixed cell sources, usually from sheep, to seed the scaffolds. This is especially important for cardiac constructs since they have to adapt to peripheral organ demands under the changing conditions of pressure and flow. The aorta, aortic valve and large arteries are able to adapt their thickness due to high collagen and elastin components and function primarily to deliver and distribute blood under high pressure to the various tissue beds.

Endothelial cells (ECs): ECs form a monolayer that constitutes the primary interface between the bloodstream and all extravascular tissue of the body. The endothelium is strategically located to serve as a sensory tissue that assesses haemodynamic conditions such as blood flow and pressure. In response to haemodynamic factors the endothelium synthesizes and secretes biologically active molecules that control smooth muscle cell tone and the vascular geometry. In recent years experimental evidence has demonstrated the importance of the cyclic stretch on ECs. i.e. cells have been seeded onto combined PHA/PHO scaffolds, following the isolation from segments of ovine carotid arteries.

Myocytes: Myocytes are muscle cells that constitute the muscular wall of the heart. Each cell beats independently at different rates and when formed in small clusters they contract spontaneously at a uniform beat rate. Electrical stimulation controls the rate of beats per minute (BPM) of these clusters. Although myocytes are rarely used to seed scaffolds for TE approaches, tests have shown that embryonic cardiomycytes from mice have a significant effect on blood vessel growth.

Fibroblasts: Fibroblasts are found in the connective tissues and secrete fibrillar pro-collagen, which forms collagen that contributes to increased tissue strength. Experiments have shown that the addition of fibroblasts to scaffolds permit secretion of collagen and matrix proteins that maintain structural integrity. In addition, several other cell sources have been proposed for use in creating TEHVs. These include cells from peripheral veins, stem cells, and circulating bone marrow-derived endothelial cells. For example, Stock et al., reported that myofibroblasts migrated into culture dishes and their number could be expanded separately.

When sufficient cell numbers were attained, these myofibroblasts were seeded onto polymer scaffolds after four days of incubation. The cells adhered to the polymer scaffolds, migrated into the pores, and secreted ECM, but the new tissue found was immature, mechanically weak, and lacked structural organization.

Smooth Muscle Cells (SMCs): Aortic valves originate from the aortic wall during embryonic development and aortic smooth muscle cells (SMCs) are often co-cultured with ECs to engineer TEHVs. However, SMCs are not found in fully developed heart valves. The media, the middle of three layers in artery walls, contains SMCs that are oriented circumferentially, within an elastin and collagen matrix. The media consists of SMCs and elastin fibers in alternating layers that form lamellar units. The elastin fibers permit distension of the artery while the collagen bundles provide tensile strength, limit distension and prevent disruption. This organization controls to the distribution and magnitude of tensile stress.

Establishment of cell lines

In order to seed fabricated polymer scaffolds, primary cell line cultures need to be established. Cell types and their sources were reviewed. For the current investigation it was decided that ovine cardiac tissue would be the most logical source as this tissue is easy to obtain and has been widely used in similar investigations. Thus, at Swinburne University cells from ovine cardiac tissue were used to seed the manually fabricated scaffold.

Experimental Procedure: Under aseptic conditions a first dissection was made at the anterior surface of the left ventricle of a lamb heart. After exposure of the left ventricle cavity, the aortic valve cusps were exposed. A surface of 5 x 5mm segment of the lamb heart valve leaflet tissue was removed from the aortic semilunar valve and 10 mm of left coronary artery. The tissue was washed three times in phosphate buffered saline (PBS) containing penicillin-streptomycin. After this procedure the tissues were divided in Petri dishes into cubes of 1 x 1 x 1 mm.

The ovine tissue-cubes were subsequently allowed to dry in the Petri dishes for approximately 1 hour. After this period 1.5 ml of Dulbecco's Modified Eagle Medium (DMEM) growth media was added. DMEM is a modification of basal medium eagle that contains four-fold concentrations of the amino acids and vitamins that support primary cultures of cells. The Petri dishes were put into a humidified incubator at 37°C in a 5% of CO2 atmosphere. The medium was changed every 24 hours. The attempts to create viable cell lines resulted in a negative outcome. Therefore, the following paragraph should be considered as a guideline for future investigators of TEHVs.

Seeding of the scaffold: After establishing viable cell cultures, cells should be seeded onto the fabricated 3D scaffolds. Seeded scaffolds should be further cultivated and assessed in the developed BR-system.

This will be the first step to engineer tissues for heart valve development. In general, basic seeding requirements include:

— High yield, to maximize cell utilization.

— High kinetic rate, to minimize the time in suspension for anchorage-dependent and shear sensitive cells.

— High and spatially uniform distribution of attached cells, for rapid and uniform tissue growth.

After one to three days the seeded PCL-scaffolds should be transferred into a BR. It is important to ensure that as many cells adhere in a confluent layer around and in the scaffold, before fixating the seeded scaffold into the BR. Although the ideal cell culture conditions are not known for each tissue, research has demonstrated that rotating the scaffold though the culture medium increases the cell attachment rate. However, other parameters such as type of cell and the type of polymer play a significant role in cell-adhesion.

Bioreactors

To grow TEHVs under conditions that mimic the in vivo environment, seeded scaffold materials should degrade over time. These environments may be created in BRs and can be divided into several categories, depending on the type of tissue required. The development of skin and cartilage but cardiac TE requires a pulsatile flow of the culture media through the construct. BRs need to incorporate flow parameters in order to simulate the in vivo cardiac environment.

Cultured cells grow more like their in vivo counterparts if the in vitro environment mimics the dynamic physical demands of the in vivo environment. In this BR cells were subjected to a unidirectional flow of media, bringing nutrients in and carrying away metabolites and other wastes. While a BR can enhance culture of cartilage cells and blood vessel cells, the results with TEHVs are particularly encouraging. Grown in a BR, tissue doubled in mechanical strength and secreted increased amount of collagen and elastin compared to cells in petri-dishes.

BRs were developed to mimic the complex in vivo environment, including the stimuli that have an impact on cell growth and differentiation. No matter what structure develops in a particular BR, the different stimuli can be divided in four basic categories:

— Growth matrix in or on which cells grow.

— The chemical and physiological composition of the medium.

— Composition of the gas in the incubator.

— Incubation temperature.

However, in order to grow viable heart valve tissue, the four above-mentioned aspects should also include mechanical stimuli that mimic in-vivo conditions. In addition to this, the BR has to meet several other requirements that include compact size, sterility, low volume, easy refreshment of the medium and access to the TEHV. The use of a pump (heat production) must not disturb the climate in the incubator. The medium should be exposed to a controlled atmosphere (diffusion of O_2, CO_2, N_2) in the incubator.

To meet these requirements, a microenvironment must be created with adequate nutrients and without accumulation of metabolites. The BR should allow testing of entire valves as well as discrete parts of it. From a haemodynamic point of view, the BR should be able to provide parameters such as transvalvular pressure gradients, flows and frequencies within certain physiological limits.

Bioreactors (BRs)

Regardless of the type of tissue cultured a particular BR, the enhancement of tissue growth in these environments must fulfil some design principles. Several studies are reviewed and presented that demonstrate how BRs can modulate 3D tissue formation in vitro. The current status of pulsatile BRs and for engineering of cardiovascular constructs. In contrast to a large number of TE studies that focus on scaffold design and in vivo tissue repair with cells and/or biomaterials. Specific BR design features may be used to improve the structure and function of engineered structures.

— One of the most basic designs of a BR is a flask system It contains 120 ml of culture medium and can contain several TE constructs depending on the size. Flasks are either operated statically or mixed, normally at 50-80 rpm. "Cardiac like tissues cultivated for one week in mixed flasks showed a DNA contents of 16% while the scell-size and the metabolic activity were similar to that in neonatal ventricles". The peripheral region of constructs was electrically excitable and could be captured over a wide range of pacing frequencies.

— The same kind of tissue have also been grown in High Aspect Ratio Vessels (HARVs). Slow Turning Lateral Vessels (STLVs) and HARVs have been used to engineer cartilage and cardiac tissues. The STLV is configured as the annular space between two concentric cylinders, the inner of which is a gas exchange membrane, whereas the HARV is cylindrical vessel with a gas exchange membrane at its base.

  Both vessels are operated in a horizontal plane at 15-40 rpm. "Cartilaginous constructs cultured in vitro for seven months in STLV's, showed a GAG fraction and equilibrium modulus that reached or exceeded values measured for native cartilage. However, other properties of seven-month constructs remained atypical; i.e. collagen fraction, cross-link concentration were only a third while dynamic stiffness was only half the value as reported in native cartilage". However, engineered tissues grown in rotating vessels were structurally and functionally superior to constructs grown in either static or mixed flasks.

— Rotating Wall Perfused Vessels were developed by the National Aeronautics and Space administration (NASA) and used to engineer cartilage in a microgravity environment of space and a control study on Earth.

— Medium was continuously re-circulated between the column and an external membrane in perfused columns. Perfused chambers were designed to allow tissue culture in the microgravity of space. The chambers can contain different volumes of culture media up to 30 ml. and hold up to five constructs. The medium is continuously circulated between the chamber and an external membrane at rates of 1-30 ml per construct a day. Although the mechanisms underlying these effects are yet to be determined, it appears that hydrodynamic forces affect cultured cells via pressure fluctuations and/or shear stress, stretching the cell membrane.

These BR functions may result in increased size and improved structure and function of engineered tissues. The BR systems described promoted mass transfer to the construct surface, but did not enhance the relatively slow diffusion of nutrients to the construct interior. This is not a major consideration for engineering cartilage, an avascular tissue with low cellularity that can be cultivated in BRs to a thickness that exceeds that of native cartilage. However, efforts to engineer tissue that has high vascularity and/or cellularity are limited and in particular, very low tissue formation has been reported for cardiac tissue.

Other groups have demonstrated advantages using BR systems that provide continuous perfusion and mechanical stimulation during cultivation, such as seeded TEHVconstructs subjected to recirculated culture media though a closed loop. In each of these cases perfusion of media led to increased tissue growth and metabolism. In addition, closed loop perfusion BRs reduce the risk of contamination during long-term cultivation. The finding that physical stimuli modulate tissue development has motivated the design of several BR systems in which growing tissues are exposed to mechanical forces. The presence of mechanical forces (externally applied or internally generated) during culture stimulated the development of the engineered tissues. This author believes that the stimulation is directly related to the physical stimuli, normally present in vivo.

Pulsatile Bioreactors (PBRs)

In the development of cardiac tissues and in particular valve leaflets, the use of mechanical forces plays a key-role. The haemodynamic function and performance of TEHVs can be improved by exposing developing tissue to physiological stresses in vitro. Therefore, a pulsatile-fluid-flow-BR (PBR) has been developed to provide physiological pressure and fluid flows. PBRs provide mechanical conditioning of constructs in the form of pulsatile culture media flow that mimics in vivo conditions.

PBRs require a sophisticated pump-system to generate cyclic fluid-flow inside the chamber. The generation of such flows is still experimental, as the ideal conditions for tissue growth are not yet established. Generated flows should be directed through the centre of the constrained cardiac construct, mimicking in vivo simulation. The following section describes and compares four investigations where a PBR was used to culture TEHVs.

This system consists of three tubes assembled into a parallel horizontal flow system where scaffolds may be secured in the tubes. The culture media is pumped through these tubes by a pulsatile pump that is controlled by a compliance chamber. In order to control the pulsatile flow, the BR was placed on a magnetic stir plate. The BR was also connected to an open medium reservoir to provide gas-exchange. The compliance chamber consisted of a 300-ml plastic reservoir that minimized the transmission of high frequency vibrations to the BRs. The flow of the culture media was applied directly through the BR at 165 BPM with 5% radial distension (strain).

PBRs used for TEHVs

Most PBRs described in the literature are able to provide pulsatile flow and consist of three major components i.e. two chambers separated by a diaphragm. The bottom chamber (1) is filled with air, and the upper chamber (2) is a dualcompartment fluid chamber. A silicone membrane (3) divides the two compartments. A sterilized TEHV construct is mounted onto a removable silicone tube (5), which is then slipped onto the fixed silicone tube (4) in the PBR.

The air chamber (1) is connected to a respirator, and when air is cyclically pumped into the lower chamber (2a), the silicone diaphragm (3) is periodically displaced, pushing fluid through the TEHV and into the perfusion chamber (2b). PBR systems like these are particularly appropriate for long term culture of cardiac neo-tissues, in particular blood vessels or heart valves. In the following paragraph the outcomes from four studies involving this type of PBR are described.

The generated fluid flow and pressures created inside the PBRs are determined by the pump-system that drives it. In the design proposed by Hoerstrup et al. pulsatile flows ranging from 50 - 2.000 ml/min and systemic pressures from 10 - 240 mmHg were generated using a respirator pump. Using this PBR, different TEHVs can be exposed to increasing levels of pulsatile flow and pressure in vitro. Following culture, the cellular response to different constructs can be evaluated with histological or biochemical techniques.

— In a study carried out by Sodian et al., the described PBR was started at a low flow condition of 140 ml/min and a systolic pressure of 10 mmHg. Over time the flow rate was increased to 350 ml/ min and the systolic pressure was set at 13 mmHg. Mixed vascular cells from adult ovine carotid artery were cultured and seeded on a tri-leaflet scaffold prior to the experiment. It was observed that by day 4 a nearly confluent cell layer was formed over the whole construct.

  Furthermore, leaflet cells oriented in the direction of flow and cells on the conduit wall formed bridges between pores. The mechanical strength of the cultured TEHV was not tested, however, ECGtests showed no pulmonary regurgitation. Furthermore, some thickening of the valve structure was observed without functional loss. Significantly more cells and collagen on the cultured constructs were detected compared to static conditioning.

— Another study conducted by Sodian et al., tested the PBR under similar conditions as described in the experiment above. Under these conditions, the fabricated TEHVs were found to open and close synchronously with pulsatile flow. Tri-leaflet scaffolds constructed from salt leached PHA were used. According to the Environmental Scanning Electron Microscopy (ESEM) analysis cells were attached to the scaffold in a nearly confluent manner and oriented themselves in the direction of flow.

Connective tissue was demonstrated within the pores of the scaffold after 4 days in the PBR. Movat staining demonstrated that the formed ECM contained both collagen and GAGs, but not elastin. DNA and 4-hydroxyproline assays demonstrated that exposure to flow increased the cell density and the amount of collagen formed in the TEHVs compared to constructs cultured in static conditions. Therefore, it was concluded that the pre-conditioning of the constructs with flow strengthens the constructs.

— Sodian et al., the BPR was used with low flow conditions of 100 ml/min for 1h. Cells were derived from a mixed population that included fibroblasts, SMCs, and ECs isolated from ovine arteries. Through a salt-leaching process, porous tri-leaflet scaffolds were constructed from PGA, PHA and P4HB. "The PHA and P4HB-constructs showed an almost confluent cell layer by day 8 and more cells attached to the PGA material than the PHA and P4HB constructs". Significantly more collagen was detected on PGA constructs when compared to the PHA & P4HB constructs.

— An experiment with more physiologically relevant pressures was performed by Hoerstrup and colleagues. TEHVs fabricated from PGA scaffolds were coated with P4HB. These constructs were placed in a PBR for 21 days. The experiment began with a flow condition of 125 ml/min and a systolic pressure of 30 mmHg. During the investigation both the flow rate and systolic pressure were gradually increased to 750 ml/min and 55 mmHg repectively. In addition, these pressures approached the maximum pressures

(56 -70 mmHg) sustained by a normal aortic heart valve in the circumferential direction at the points of leaflet attachment to the conduit wall. In contrast to findings that TEHVs were fragile and disintegrating after 14 days of culturing in static conditions, TEHVs cultured for a similar time period in the PBR were intact, pliable, and competent in closure.

Following a pre-conditioning regimen, the valve substitutes engineered by Hoerstrup et al. were implanted into the supravalvular position of pulmonary arteries in lambs. No rejection was detected as the implanted constructs were engineered from the samess lambs from which the cells were initially harvested. Prior to implantation, each construct was subjected to high-pressure testing, with pressures > 150 mmHg for 1h. In vivo valve function was evaluated via a Doppler echocardiogram.

The TEHVs functioned for up to twenty weeks without stenosis, thrombosis, or aneurysm formation, but moderate pulmonary regurgitation and inflammation were observed between 16 - 20 weeks. Histological analysis revealed a patchy EC-layer and incomplete polymer degradation. The mechanical strength of the TEHV conduit walls was also determined using an Instron mechanical testing apparatus. The stress-strain curve the TEHV resembled that of a native pulmonary artery. Several studies have been performed with cultured TEHVs under pulsatile flows within the last four years. It was noted that all constructs in these studies were seeded with cells under static conditions prior to culture.

Chapter 8

Confocal Microscopy of Live Cells

Microcinematography and, later, video microscopy have provided great insight into biological phenomena. One limitation, however, has been the difficulty of imaging in three dimensions. In many cases, observations have been made on cultured cells that are thin to start with or tissue preparations that have been sectioned. The development of the first beam-scanning confocal microscope was motivated by the goal of making observations in the tissues of living organisms.

The optical sectioning capability of the confocal or multiphoton (MP) microscope allows one to make thin-slice views in intact cells or even intact animals. Confocal microscopes are now fairly common, and because they employ non-ionizing radiation, they are increasingly being used to study living cells and tissue preparations. First, the experimenter must do no harm. Arguably, the main obstacle in living cells microscopy is not "getting an image" but doing so without upsetting the cell.

To be useful, the study must be carried out on a biological system that retains normal function and can be subjected to controlled conditions while on the stage of the microscope. Often, environmental variables such as temperature, CO_2, or pH must be regulated and/or an efficient directional perfusion system must be used. Unfortunately, the difficulty of keeping cells alive and functioning on the microscope discourages many researchers. Other difficulties are more specific to confocal fluorescence microscopy. All studies with fluorescence benefit from collecting as much of the emitted fluorescent light as is possible, but this is particularly important for studies of living cells because photodynamic damage and consequent alteration in normal cell behavior is a very real possibility.

Therefore, optimizing microscope photon collection efficiency is crucial for successful confocal microscopy of living cells. Another difference between living and fixed cell studies is the element of time. All living processes have an inherent time course, and the imaging system must produce images at the appropriate rate to show the changes involved. The amount of light necessary to obtain the data must be apportioned over time so that enough images can be obtained to describe the process under investigation without damaging the cells.

Although early confocal microscopes had a relatively slow scan speed, newer technology now permits very rapid image collection to explore spatially and temporally dynamic biological processes. In single-beam scanning systems, the field of view often is reduced to achieve higher imaging speeds. Lastly, the fluorescent probes used in studies of living cells must not impair normal cell function. Immunofluorescence, which has been used so successfully to localize molecules in fixed cells, has not been practical in living cells.

However, there are now many commercially available fluorescent probes for structural and physiological studies of cells and tissues. Of even more importance, the 'green revolution' based on the green fluorescent protein (GFP) has changed the landscape and is ushering in an exciting period of biological imaging of proteins in living cells, and of various cell types in living, intact tissue preparations. Although confocal microscopy of living cells is difficult, its usefulness was demonstrated over 15 years ago in two pioneering studies.

Cornell-Bell et al. used confocal microscopy to make a major discovery: the existence of glutamate-stimulated, transcellular Ca^{2+} waves in astroglia. In the same year, confocal microscopy was used to characterize developmental changes in an intact animal by imaging neuronal axons and their growth cones in the developing brain of a tadpole. Ever since these pioneering studies, there has been an increasing use of confocal microscopy to study dynamic processes in an array of diverse biological preparations.

While live-cell applications of confocal imaging have expanded significantly over the past decade, multiphoton (MP) imaging is poised to make a similar impact on live -cell and tissue studies in the next decade. There are trade-offs, however. MP imaging can be useful especially. for very deep penetration in tissues (>100μm) where non-descanned detection increases signal substantially, but from thinner specimen, the actual damage/excitation may be greater for MP than for single photon confocal imaging. Moreover, the cost differential is such that one could have 2-3 graduate students working away on disk scanners or simpler beam scanning confocal units for every one on a MP unit.

LIVE -CELL CONFOCAL IMAGING TECHNIQUES

Although live -cell imaging often involves time-lapse microscopy to monitor cell movements, modern approaches are extending these observations well beyond simply making movies of cell structure. Increasingly, time-lapse imaging is being integrated with specialized techniques for monitoring, measuring, and perturbing dynamic activities of cells and subcellular structures. Some major techniques available for studying the dynamic organization of molecules and cells in live biological specimen.

Time-lapse Fluorescence Imaging

Time-lapse fluorescence imaging involves repeated imaging of a labeled specimen at defined time points, thereby permitting studies on the dynamic distribution of fluorescently labeled components in living systems. Imaging can be performed in one, two, or three spatial dimensions: 1-D imaging involves rapid and repeated imaging of single scan lines; 2-D imaging involves repeated imaging of single focal planes; and 3-D imaging involves repeated imaging of multiple focal planes in thick specimen.

The time intervals for sequential image collection can range from sub-second to days or even months. Many small-molecule, vital fluorescent probes that give highly specific cellular or subcellular patterns of labeling are now available. In addition, GFP or GFP-related proteins are now routinely fused to other proteins of interest, and the inherent brightness and photostability of many of these fluorescent proteins make them well suited for the repeated imaging needed for time-lapse studies. Together, these fluorescent probes are affording a seemingly limitless array of possibilities for imaging molecular components in live cells.

Multi-channel Time-lapse Fluorescence Imaging

The plethora of excellent vital fluorescent labels with varying spectral characteristics (including spectral variants of GFP) allows multi-label experiments to visualize the relative distribution of several different cell or tissue components simultaneously.

Advances in imaging technology have facilitated automated collection of more than one fluorescent channel with improved ability to maximize signal collection and to separate partially overlapping signals. In addition to studies using multiple fluorescent tags, multi-channel data collection permits ratiometric imaging of single probes whose spectral properties change depending on ionic conditions, such as the Ca^{2+} sensitive physiological indicator, indo-1.

Spectral Imaging and Linear Unmixing

Increasingly, experiments are incorporating multiple fluorescent probes within single cells or tissues to define the differential distribution of more than one labeled structure or molecular species. Such multi-color or multi-spectral imaging experiments require adequate separation of the fluorescent emissions, and this is especially problematic when the spectra are substantially overlapping. Spectral imaging utilizes hardware to separate the emitted light into itsspectral components.

Linear unmixing is a computational process related to deconvolution that uses the spectra of each dye as though it were a point-spread function of fixed location to "unmix" the component signals. Although together, these analytical tools can be used to discriminate distinct fluorophores with highly overlapping spectra, they do so at the cost of requiring that significantly more photons be detected from each pixel.

Fluorescence Recovery After Photobleaching (FRAP)

Fluorescence recovery after photobleaching (FRAP), also known as fluorescence photobleaching recovery (FPR), is a technique for defining the diffusion properties of a population of fluorescently labeled molecules. Typically, a spot or line of intense illumination is used to bleach a portion of a fluorescent cell, and the recovery of fluorescent signal back into the bleached area from adjacent areas is monitored over time.

Although this technique can yield quantitative information on the diffusion coefficient, mobile fraction, and binding/dissociation of a protein, care has to be taken not to use so much power in the bleach beam that the cellular structure is disrupted. Quantitative assessments of FRAP data, which can be confounded by uncertainties in the experimental and biological parameters in living cells, may benefit from computer simulations.

Fluorescence Loss in Photobleaching (FLIP)

This technique utilizes repeated photobleaching in an attempt to bleach all fluorophore within a given cellular compartment. Thus, FLIP can be used to assess the continuity of membrane bounded compartments (e.g., ER or Golgi Apparatus) and to define the diffusional properties of components within, or on the surface of, these compartments.

Fluorescence Resonance Energy Transfer (FRET)

Fluorescence resonance energy transfer (FRET) is a technique for defining interactions between two molecular species tagged with different fluorophores. It takes advantage of the fact that the emission energy of a fluorescent "donor" can be absorbed by (i.e., transferred to) an "acceptor" fluorophore when these fluorophores are in nanometer proximity and have overlapping spectra.

Fluorescence Lifetime Imaging (FLIM)

This technique measures the lifetime of the excited state of a fluorophore. Each fluorescent dye has a characteristic "lifetime" in the excited state (usually 1-20 nanoseconds), and detection of this lifetime can be used to distinguish different dyes in samples labeled with multiple dyes. FLIM can be utilized in conjunction with FRET analysis because the lifetime of the donor fluorophore is shortened by FRET. In fact, FLIM can improve the measurement during FRET analysis because the fluorescence lifetime is independent of the fluorophore concentration and excitation energy. However, the lifetime can be modulated by environmental considerations, and this change can be used to measure changes in the concentration of certain ions.

Fluorescence Correlation Spectroscopy (FCS)

Fluorescence correlation spectroscopy (FCS) measures

spontaneous fluorescence intensity fluctuations in a stationary microscopic detection volume (about 1 femptoliter). Such intensity fluctuations represent changes in the number or quantum yield of fluorescent molecules in the detection volume. By analyzing these fluctuations statistically, FCS can provide information on equilibrium concentrations, reaction kinetics, and diffusion rates of fluorescently tagged molecules. An advantage of this approach is the ability to measure the mobility of molecules down to the single molecule level and to do so using a light dose orders of magnitude lower than used for FRAP.

Fluorescence Speckle Microscopy (FSM)

The dynamic growth and movement of fluorescently labeled structures can be difficult to analyze when these structures are densely packed and overlapping within living cells. Fluorescent speckle microscopy (FSM) is a technique compatible with widefield or confocal microscopy that uses a very low concentration of fluorescently labeled subunits to reduce out-of-focus fluorescence and improve visibility of labeled structures and their dynamics in thick regions of living cells.

This is accomplished by labeling only a fraction of the entire structure of interest. In that sense, it is akin to performing FCS over an entire field of view, albeit with more focus on spatial patterns than on quantitative temporal analysis. FSM has been especially useful for defining the movement and polymerization/depolymerization of polymeric cytoskeletal elements, such as actin and microtubules, in motile cells.

Photo-uncaging/Photo-activation

Photo-uncaging is a light-induced process of releasing a 'caged' molecule from a caging group to produce an active molecule. A variety of caged molecules have been synthesized and used experimentally, but in some instances cages have been used to mask a fluorophore, inducing a non-fluorescent state. Excitation light of ~350nm is used to break photolabile bonds between the caging group and fluorophore, thereby uncaging the fluorophore and yielding a fluorescent molecule. A related technique utilizes genetically encoded, photo-activatable fluorescent proteins, of which there are currently about a dozen.

Two examples include a photo-activatable (PA) form of GFP, called PA-GFP, which shows a 100-fold increase in fluorescence following irradiation at 413 nm, and Kaede, which shows a 2,000-fold increase following irradiation at 405 nm. An extension of the photo-activation approach, termed reversible protein highlighting, has been developed. This involves reversible, light-induced conversion of a coral protein, Dronpa, between fluorescent and non-fluorescent states.

One study used this approach to monitor fast protein dynamics in and out of cell nuclei. Thus, photo-uncaging and photo-activation are complementary to FRAP and can be used in conjunction with time-lapse imaging to mark and follow a population of molecules in order to study their kinetic properties within living cells.

Optical Tweezers/laser Trapping

Optical tweezers, or single beam laser trap, uses the 'radiation pressure' of a stream of photons emitted from an infrared laser to "trap" small objects (often a protein-coated bead) and to move them around. This technique has been especially useful for quantifying forces generated by motor protein movement or the strength of adhesions mediated by cell adhesion molecules. Although "laser tweezers" often are used in widefield imaging systems, they also have been incorporated into confocal and MP imaging systems.

Physiological Fluorescence Imaging

The availability of fluorescent physiological indicators extends live -cell confocal and MP imaging studies beyond structural aspects to study cell and tissue physiology. Calcium indicators have been the most commonly used physiological probes because calcium is a central signal transduction molecule and in many cell preparations the calcium-sensitive probes give robust signals. These signals often are temporally resolvable in full field scans as calcium transients that persist for several seconds. Fast scanning systems, or line-scanning mode in laser scanning systems, have been used to resolve more rapid calcium events.

Although non-ratiometric, visible wavelength calcium indicators (e.g., fluo-3, calcium green) have been more widely used in confocal applications, some studies have employed UV excited ratiometric calcium indicators, such as indo-1. In addition to calcium indicators, other fluorescent physiological probes are useful for reporting various ions including sodium, magnesium, potassium, and chloride, pH, heavy metals such as zinc, and membrane potential, to name a few.

Although many of these probes are small molecules, genetic (GFP-based) probes have been developed and are being incorporated into transgenic animals. In combination with state-of-the-art confocal and MP imaging systems, these probes will increasingly permit detailed spatio-temporal analyses of physiological processes in intact tissues and organisms.

Combining Fluorescence

Although advancements in fluorescence imaging technology coupled with the availability of a multitude of vital fluorescent probes have combined to make fluorescence the method of choice for most high resolution studies of living cells, it is sometimes advantageous to combine fluorescence imaging with other imaging modalities. For example, differential interference contrast (DIC) microscopy can be used in conjunction with scanning laser confocal microscopy to simultaneously monitor the whole cell in DIC mode while imaging the phagocytic uptake of fluorescent microspheres or the distribution of fluorescently-tagged proteins and molecules within these cells.

Although it is difficult to perform DIC and epi-fluorescence imaging both simultaneously and optimally in widefield microscopy, it is somewhat easier to ensure that the fluorescence signal is not subjected to the light loss that occurs in the analyzer used as part of the DIC system if one uses a single-beam confocal.

Thus, the DIC image can be collected from a fluorescently labeled specimen using transmitted light that would otherwise be wasted. Recently, differential phase contrast (DPC) has been implemented in a scanning laser microscope system, and this may offer additional capabilities where DIC optics are unsuitable. Notably for live-cell imaging, DPC reportedly needs 20 times less laser power at the specimen than DIC.

When performing a live-cell confocal imaging experiment or observation, have the major factors are to (1) label the preparation in order to clearly visualize the biological component of interest, (2) maintain the preparation in a condition that will support normal cell or tissue health, and (3) image the specimen with sufficient spatial and temporal resolution in a way that does not perturb or compromise it.

Living Cells and Tissue Preparations

In vitro preparations

Specimen maintenance is a very important part of any live imaging study and usually requires both mechanical ingenuity and insight into the biology of the cell or tissue under study. The specimen chamber must keep the cells or tissues healthy and functioning normally for the duration of the experiment while allowing access to the microscope objective. This can be particularly difficult when high-numerical-aperturc (NA) oil- or water-immersion lenses are used. In many cases, there must also be a controlled and efficient way to introduce a reagent to perturb a particular cellular process.

Other important factors are simplicity, reliability, and low cost. At any rate, it is advisable to monitor the conditions within the imaging chamber carefully. It may be helpful to use microprobes that can detect pH, O_2, and CO_2 . The early closed perfusion chambers designed by Dvorak and Stotler and later by Vesely et al were inexpensive and permitted high-resolution transmitted light observation. They relied on an external heater that warmed the entire stage area for temperature control.

Setups for different cells vary widely. Mammalian cells probably pose the greatest problems. McKenna and Wang's article is a general introduction to the problems associated with keeping such cells alive and functioning on the microscope stage. Strange and Spring describe their setup for imaging renal tubule cells where temperature, pH, and CO_2 are controlled. They provide a detailed account of the problems of establishing laminar flow perfusion systems, temperature regulation, and maintenance of pH by CO_2 buffering.

Somewhat later, Delbridge et al. describe a sophisticated, open-chamber superfusion system permitting programmed changes of media, precision control of media surface height, and temperature regulation between 4°C and 70°C using a Peltier device to control the perfusate temperature. Myrdal and Foster used a temperature-stabilized liquid passing through a small coil suspended in media filling a plastic NUNC chamber to provide temperature control for confocal observations of the penetration of fluorescent antibodies into solid tumor spheroids.

An automatic system maintained fluid level and bathed the area in CO_2 but special precautions were required to prevent drift of the confocal focus plane during long time-lapse sequences. Chambers have even been built for the microscopic observation of cells as they are being either frozen or thawed in the presence of media that could be changed during the process. In this case, computer-controlled pumps deliver temperature-controlled nitrogen gas at between -120°C and 100°C to special ports connected to a temperature cell (-55°C to 60°C) that forms the upper boundary of the perfusion chamber.

Other ports carry either the perfusate or a separate nucleating agent tò the cell chamber itself. More recently, a specialized in vitro cell culture system has been developed to maintain mammalian neuronal cells for over a year! There are several companies that provide ready-made microscope stage chambers, temperaturecontrol units, automated perfusion systems, and a variety of related accessories.

In vivo preparations

The ultimate goal of many research programs is to understand the normal (or abnormal) structure and function of molecules, cells, and tissues in vivo, that is, in the living organism functioning within its native environment. There has been some remarkable progress recently on extending high resolution confocal and MP imaging in this direction, especially in preparations that are essentially translucent Several model organisms, including zebrafish, frog, fruit fly, leech, and worm, have emerged as excellent preparations for cellular and molecular imaging studies spanning a variety of biological questions.

As an example, studies in the zebrafish have been carried out on the structural development of vasculature, cell division, neuronal migration, axonal pathfinding, synapse formation, and synaptic plasticity, to name a few. Physiological studies in zebrafish have included, for example, imaging intracellular calcium during gastrulation, in the intact spinal cord, and in brain. Each of these biological preparations embodies its own unique set of specimen mounting and maintenance challenges. Indeed, it is sometimes necessary to anesthetize the preparation to prevent it from crawling or swimming away during the imaging session.

Perhaps the most difficult conditions involve imaging in a living mammal, an undertaking for which the confocal or MP microscope enjoys the twin advantages of epi-illumination and optical sectioning that make it possible to view solid tissues without mechanical disruption.

Confocal microscopy has long been an important tool for in vivo imaging of eye tissues non-invasively. In terms of imaging interior tissues, early studies described methods for examining microcirculation of the brain cortex in anesthetized rats or changes in kidney tubules during ischemia. Confocal microscopy also has been used to image leukocyte-endothelium interactions during infections through closed cranial windows.

More recently, MP has been used to image live mammalian brain tissues in vivo, either through a cranial window, fiber optic coupled devices, or directly through the intact but thinned skul. Dual-channel MP imaging also has been used to image other tissues in vivo, including lymphoid organs. It is generally accepted that MP imaging is superior to single photon confocal for these in vivo imaging studies.

Fluorescent Probes

Except in those cases where an adequate image can be derived from either the backscattered-light signal or from autofluorescence, confocal microscopy of living cells is dependent on the properties and availability of suitable fluorescent probes. In addition to binding specifically to what one is interested in studying, the fluorescent probe should produce a strong signal and be both slow to bleach and nontoxic. Many dyes are useful when introduced to the medium surrounding cells to be labeled.

Some of the classic and most commonly used cell stains include DiI for labeling the plasma membrane, $DiOC_6(3)$ for labeling internal membranes, NBD-ceramide and bodipy-ceramide which label the Golgi apparatus, rhodamine 123 which labels mitochondria, potential sensitive dyes, and FM 1-43 which is used to follow plasma membrane turnover and vesicular release. Fluorescent ion indicators such as Fluo-3 can either be microinjected or added to the media in a cellpermeant acetoxymethylester form that becomes trapped inside the cell after being cleaved by intracellular esterases.

Photodynamic Damage

Once the cells are labeled and on the microscope, one is faced with the challenge of collecting data without compromising the cell or bleaching the label. In practice, the major problem is light-induced damage. Fluorescent molecules in their excited state react with molecular oxygen to produce free radicals that can then damage cellular components and compromise cell health. In addition, several studies suggest that components of standard culture media might also contribute to light-induced adverse effects on cultured cells.

Some early studies indicated a phototoxic effect of N-2-hydroxyethylpiperazine-N'-2-ethanesulfonic acid (HEPES) containing media on cells under some circumstances. It seems possible that this effect might be more directly related to inadequate levels of bicarbonate. Riboflavin/vitamin B2 and the essential amino acid tryptophan may mediate phototoxic effects. Whether these effects occur under typical confocal imaging conditions is unknown, but many of the photoeffects are eliminated by anti-oxidants, so it seems advisable to maintain anti-oxidants in the specimen chamber and to use photons with great efficiency.

Improving photon efficiency

There are several strategies to minimize the amount of excitation light required to collect data. Briefly, higher-NA objective lenses collect more of the fluorescent emission. For a given lens, there is also a theoretical optimal setting of the zoom magnification that best matches the resolution required to the allowable dose. When the focus plane is more than 10μm from the coverslip, water-immersion lenses should be used to avoid the signal loss caused by spherical aberration when using an oil lens. Another way to reduce light damage is to minimize the duration of the light exposure during the experimental setup.

For instance, one should try to focus as rapidly as possible and turn off the light source as soon as the focus range has been chosen. In addition, in single beam scanning systems, make sure that your scanner is set up to blank the laser beam during scan retrace. Otherwise, areas on both sides of the imaged area will receive a very high light exposure as the beam slows down to change direction. Finally, photon efficiency can be maximized by using the best mirrors, the correct pinhole size for the resolution required (in x, y, and z), and photodetectors that yield the highest quantum efficiency at the wavelength of the signal.

Anti-oxidants

One can also reduce photodynamic damage by adding anti-oxidants to the medium. Oxyrase is an enzyme additive used to deplete oxygen in order to grow anaerobic bacteria. It has been used at 0.3 unit/ml to reduce photodynamic damage during observations of mitosis. Another approach is to include ascorbic acid in the medium. This reducing agent is typically used at 0.1-1.0 mg/ml but has been used at up to 3 mg/ml. A confocal study of calcium transients in isolated chrondrocytes reported a relationship between laser intensity and the frequency of Ca^{2+} oscillations and cell viability: Ca^{2+} events were more frequent and cell viability was decreased with higher laser intensity. Treatment with ascorbic acid reduced the Ca^{2+} events and improved cell viability.

On-line Confocal Community

Confocal microscopy of living cells is an area of active research where individuals are constantly developing new techniques and approaches. One way to keep up with current practice is to join about 1,600 others who subscribe to the Confocal e-mail listserver. You will then begin to receive messages from other microscopists. Recent topics have included discussions on such diverse issues as autofluorescence problems, glass-bottomed culture chambers, damage to live cells during FRAP experiments, and announcements of confocal workshops.

Convenient Test Specimen

Onion epithelium (Allium cepa) is a simple preparation that can be used as a convenient test specimen for confocal microscopy of living cells. First, a small square of a layer is cut out using a razor blade. A forceps is used to peel off the thin epithelium on the inner surface of the onion layer. The epithelium is then put onto a microscope slide, covered with a drop or two of staining solution containing DiOC6(3), a marker of mitochondria and endoplasmic reticulum, and coverslipped.

The stock solution of DiOC6(3) (0.5 mg/ml in ethanol) can be kept indefinitely if protected from light in a scintillation vial. The staining solution is a 1:1000 dilution in water on the day of the experiment. The center of these cells is usually occupied by a large vacuole, and the ER and mitochondria are located in a thin cytoplasmic region near the plasma membrane. Motion of the ER is relatively quick and easily detected in consecutive 1-sec scans.

Microscopy of living cells has become an technique of major importance in cell biology: it can be used to tell us where molecules are located, when they become localized, how fast they are moving, with which molecules they are interacting, and how long they stay attached to these molecules. All these properties can be observed in the natural environment of the living cell. The major limiting factor in live-cell imaging is phototoxic effect of light used for the observation of the cell. Here address some practical issues of phototoxicity based on our experience in imaging chromatin dynamics in living cells.

Phototoxicity

A large number of photochemical reactions are responsible for the phototoxic effect of light. Light can be absorbed by cellular components and induce chemical alterations in the molecular structure. For example, UV-light is absorbed by DNA, directly inducing DNA-damage. Working with visible light, the direct photodamage is negligible. In fluorescently labelled cells, the main source of photodamage is the production of reactive oxygen species (ROS) including singlet oxygen ($^1O_2$), superoxide ($\cdot O_2^-$), hydroxyl radical (HO·), and various peroxides.

These activated oxygen species react with a large variety of easily oxidizable cellular components, such as proteins, nucleic acids, and membrane lipids. Singlet oxygen is responsible for much of the physiological damage caused by reactive oxygen species. For the production of singlet oxygen the fluorescent label acts as a photosynthesizer in a photochemical reaction where dioxygen ($^3O_2$) converts into singlet oxygen ($^1O_2$). Singlet oxygen mainly modifies nucleic acid through selective (oxidative) reaction with deoxyguanosine into 8-oxo-7,8-dihydro-2'-deoxyguanonine. Proteins and lipids also will be damaged by ROS. Phototoxicity likely depends on several variables:

— *The photochemical properties of the fluorescent molecule*: Some molecules induce more phototoxicity than others, depending on the lifetime of their triplet-state. For photodynamic therapy (PDT) dedicated molecules called photosensitizers, have been designed in order to induce a maximum damage in tissue for the treatment of cancer (e.g. halogenated fluorescein is much more toxic than fluorescein). Another property that influences the phototoxicity of a molecule is the local environment of the molecule. The active fluorophore of a GFP molecule is positioned on the inside of the protein, within the barrel structure (the "b-can"). Probably this hydrophobic, proteinenvironment contributes to the relatively low phototoxicity of GFP compared with 'naked' fluorophores such as fluorescein or rhodamine.

— *The subcellular location of the fluorescent molecule*: When fluorescent molecules are situated close to DNA, the damaging effect of singlet oxygen is more pronounced. Despite several DNA-repair mechanisms, the cell will not continue its cell cycle (arrest) and may even die if there is too much DNA damage. Therefore, fluorophores in the cytoplasm seem to induce less phototoxicity than fluorophores in the nucleus.

— *The concentration of fluorophore*: It is clear that there is a relationship between the local concentration of fluorophore and the level of phototoxicity. A linear relationship between fluorophore concentration and toxicity, although this has not been assessed directly and is complicated by the fact that if there is more dye, one need use less excitation.

— *The excitation intensity*: Fluorescent cells in a dark incubator are quite happy for weeks as long they are maintained in the dark. As the word 'phototoxicity' implies, photons are needed to induce toxicity in fluorescently labeled specimen. A linear relationship between excitation light dose and toxicity, although the temporal regimen of the excitation may be important to how cells handle the accumulation of phototoxic biproducts. Phototoxicity is dependent on the wavelength of light in the sense that the wavelength of the 'toxic' excitation light matches the excitation curve of the fluorophore. In other words, it is the excited fluorophore that is toxic. Koenig also found that, with 2-photon excitation, the damage is proportional to the number of molecular excitations.

There is no clear evidence for differences in phototoxicity between green, red, or far-red fluorophores. In principle, excited Cy5 can be as toxic as excited FITC. However, the wavelength of excitation light can be a factor when imaging in thick specimen because stronger incident illumination is needed for comparable excitation of shorter wavelength fluorophores due to increase tissue scatter at shorter wavelengths.

Reduction of photo-toxicity

Phototoxicity is a serious limitation of their observations of living cells. When you do not look at a cell it is alive, but the moment you start to observe how it lives, it is killed by the light used to observe it. Obtaining acceptable time series of living cells by carefully optimising all steps in the imaging process in an effort to achieve (i) maximal signal-to-noise ratio, (ii) maximal spatial and temporal resolution, and (iii) minimal phototoxic effects.

Specifically, phototoxicity has been minimized by (i) using radical scavengers (e.g., trolox) in the culture medium, and (ii) using culture medium without phenol-red. Most important of all, however, is minimizing the total excitation light dose. The excitation light dose is the product of the light intensity and the exposure time. Decreasing either the excitation intensity or the excitation dose implies a loss of fluorescent signal. It is inevitable that a reduction of light dose puts a limitation on the S/N ratio and the spatial and temporal resolution. The art of live-cell microscopy is finding the balance between image quality and cell vitality.

Improving image quality in low-dose microscopy

The cell is in late telophase at the start of the imaging and proceeds into interphase during the movie. This movie shows data from a study on the dynamics of chromatin during decondensation. Under these conditions the total light dose was approximately10 J cm^{-2}.

Reducing the total light dose during an experiment requires that the number of 3D images in the sequence (temporal sampling rate) is low. Because of this limited sampling rate, live-cell movies are usually under-sampled in time according to the Nyquist criterion. Such movies often show cells that nervously move from one place to another and sometimes suddenly rotate. An image processing procedure to correct for all the movements (translation and rotation) of the cell. For each 3D image of the time sequence, a translation and rotation transform vector was calculated in order to obtain a best fit with the previous image in the sequence.

After a series of such transformations, a new movie was produced showing a stable cell that does not move or rotate. Only internal movements are visible. After this correction procedure, here applied a simple Gaussian spatial filter to reduce noise in the image. A temporal filter by adding to each voxel of the 3D image at each time-point the value for that voxel in the previous and subsequent image multiplied by an intensity factor of 0.5.

The success of live -cell microscopy is very much dependant on minimizing or avoiding any toxic effect of light on the biological system under observation. A certain dose of light may induce serious DNA damage that may arrest the cell cycle, whereas the diffusion coefficient of a certain protein is not influenced at all at the same dose. This dose was found to be phototoxic in other experiments using fluorescein instead of GFP, and it was found necessary to drop the laser power to 50 nW. These power levels are far lower than (i.e., <1% of) those commonly used in confocal microscopy, a circumstance facilitated at least in part by the fact that the chromosomes are quite heavily stained.

A collective experiences indicate that the effect of phototoxicity depends on the cell type, the stage of the cell cycle, the fluorophore, the observed biological process, and many other experimental conditions. There is no general guideline for the maximum allowable laser power: it must be assessed empirically for each experimental condition. As a general rule, however, images of living cells are almost always more noisy than images of fixed preparations because the incident illumination intensity needs to be kept to a minimum to maintain cell viability. A noisy image in which one can see what is absolutely essential is of more use than a "better" image of a damaged cell.

Keep in mind that not all types of damage are equally easy to detect. Damage may disturb a monitored process, it may interfere with cell division, or it may cause the cell to bleb and pop. The experimentalist BEGIN by assuming that "to observe is to disturb." Measure the power level coming out of the objective. Try your experiment again using twice the power, and again using half the power. Make sure that you can explain any 'behavioral' differences between these runs.

The example above serves to illustrate that confocal microscopy is an important tool for studying dynamic subcellular processes in live, isolated cells. Many biologists are also interested in understanding dynamic structural and functional aspects of cells within the context of a natural tissue environment. As noted above, confocal and multiphoton imaging have been applied to intact, normally functioning systems such as the eye, skin, or kidney. Some recent studies have even extended these observations beyond superficial tissues to deep tissues of the brain.

However, some tissues are much less accessible, or it may be of interest to be able to experimentally perturb or control the system under study. For these purposes, the in vitro tissue slice has been an important experimental preparation. Smith et al. were among the first to show the feasibility of imaging the structure and physiology of living mammalian brain tissue slices at high resolution using fluorescence confocal microscopy. Since then many confocal studies of both structural and physiological dynamics of cells in tissue slices have appeared, and it seems that interest in imaging in vitro tissues is continuing to grow.

Here some of the most common problems, challenges, and limitations inherent in confocal studies of live tissue slices. Some of the major problems encountered when imaging fluorescently labeled cells in live tissue slices are:

— Attaining a suitable level and specificity of staining.

— Maintaining cell/tissue health: pH, temperature, oxygen, etc.

— Keeping cells in focus: can be an immense problem when following cells over long periods of time:

  - movement of the microscope stage, especially when stage heaters are used.
  - movement of the tissue: apparent movement that is really caused by movement of the focal plane within the specimen; natural movement of whole organisms or those caused by heartbeat, etc.
  - movement of cells within the tissue, e.g., cell migration, extension/retraction of cell processes.
  - movement related to experimental procedures, e.g., stimulus-induced osmotic changes.

— Attaining a useful image with a high signal-to-noise ratio of cells deep within tissue:

  - imaging away from damaged tissue surfaces.
  - light scatter by the tissue.
  - the problem of spherical aberration.

— Handling data: viewing, storing, retrieving, and analyzing 4D data sets:

  - short-term: monitoring experiments on the fly; adjusting focus.
  - long-term: accessibility and security of archived data.

The dynamic behavior of a type of brain cell, termed microglia, following brain tissue injury. These cells undergo a dramatic transformation ("activation") from a resting, ramified form to an amoeboid-like form within a few hours after traumatic tissue injury. Activation of microglia is triggered by signals from injured cells (including neurons), and this mobilizes microglia to engage neighboring dead and dying cells. Naturally, these events are best studied in the context of a complex tissue environment containing the native arrangement of tissue components; thus time-lapse confocal microscopy is well suited to examine these events. The general approach have taken is to label the cell surface of microglia with fluorescent probes and, subsequently, to follow the dynamic movements of these cells, as well as their interactions with other labeled cells, within live tissue slices continuously over periods of time up to 28 hr.

CNS tissue slices

A useful method of preparing and maintaining live brain tissue slices for microscopy is based on the organotypic (roller-tube) culture technique of Gähwiler or the static filter culture technique of Stoppini et al.. Briefly, these techniques involve rapidly removing the tissue of interest (in this case, neonatal rat or mouse hippocampus), then slicing the tissue with a manual tissue chopper (Stoelting, Chicago, IL) at a thickness of 300-400 μm.

Others have used a vibratome or custom-built instruments akin to an egg slicer. In the case of the roller tube technique, the tissue slices are secured to an alcohol-cleaned coverslip (11 x 22 mm) with a mixture of chicken plasma (10 μl; Cocalico) and bovine thrombin (10 μl;Sigma). Collagen gels and Cell-Tak have also been used successfully to attach slices. In the case of the plasma clot, the slices are adherent within about 10 min, at which point the coverslips are placed in a test tube with 1 ml of HEPESbuffered culture media containing 25% serum.

The tubes are kept in a warm box (37°C) and rotated at 12 rph in a roller drum tilted at 5° to the horizontal. In the case of the static filter cultures, brain slices are placed on porous cell culture inserts in 6-well plates containing ~1ml of bicarbonate-buffered culture media per well. The filter cultures are maintained at 36°C in a 5% CO_2 incubator. In either case, these "organotypic" culture methods provide a means for maintaining tissue slices in vitro for up to several weeks.

Fluorescent staining

Often it is most useful to label only a small percentage of the total number of cells within a tissue volume, and in certain cases it is desirable to label only an identified subset of cells. Vital fluorescent probes must be non-toxic and resistant to photobleaching. In the case of microglia, there are several commercially available fluorescent conjugates of a highly selective, non-toxic lectin (IB4) derived from Griffonia simplicifolia seeds.

This lectin has an exclusive affinity for a-D-galactosyl sugar residues on glycoproteins and glycolipids. In mammalian brain tissues, IB4 labels only microglia and endothelial cells lining blood vessels and capillaries. The simultaneous labeling of microglial cell populations and blood vessels has revealed novel, dynamic interactions between these structures.

Incubation of brain slices or slice cultures for 1 hr in IB4-containing medium (5μg/ml) is sufficient for robust labeling of microglia up to ~50μm deep within tissues. Nuclei of live or dead cells. One of a variety of fluorescent DNA-binding dyes is used to label live or dead cell nuclei in brain tissue slices. To visualize the nuclei of live cells, here use one of the membrane-permeant dyes that have spectra in the far-red, SYTO59 or SYTO61.

Tissue health on the Microscope stage

Image data obtained from compromised tissue is useless at best, and deceiving at worst. For example, CNS slice physiologists have long known that oxygen deprivation can have severe effects on synaptic activity, although CNS tissues from developing animals seem to have a fairly high resistance to hypoxia. It is not always easy to assess the health of living tissue on the microscope stage, but in the case of dynamic processes such as cell division or cell migration, one would at least expect that the cells perform these activities at rates near that expected based on other methods of determination.

Also, one should become suspicious if the rate of activity consistently increases or decreases over the imaging session. For example, exposure of fluorescently labeled axons to high light levels can reduce the rate of extension or cause retraction. In contrast, high light levels can produce a long-lasting increase in the frequency of Ca^{2+} spikes in Fluo-3-labeled astrocytes in cultured brain slices. In many cases, there will not be a useful benchmark for determining phototoxic effects, but consistent changes during imaging will serve to warn the concerned microscopist.

It may be worth sacrificing a few well-labeled preps to determine if different imaging protocols, such as lower light levels or longer time intervals between images, significantly alter the biological activity under study. Requirements for maintaining healthy tissue during imaging dictate specimen chamber design. A closed specimen chamber has the advantage of preventing evaporation during long experiments and stabilizing temperature fluctuations caused by this evaporation.

Microglia in tissue slices maintained in a closed chamber (volume ~1 ml) with HEPES-buffered culture medium remain viable and vigorous for about 6 hr, after which point the chamber medium acidifies and cell motility declines. However, when the old chamber medium is exchanged with fresh medium, the cells jump back to life again. This crude method of periodic medium exchange has supported continuous observation of DiI-labeled migrating cells in tissue slices on the microscope stage for as long as 45 hr.

However, when using this approach, one runs the risk of mechanically disturbing the chamber or inducing a temperature change and thereby causing a jump in focus. A more elegant method for medium exchange and introduction of reagents involves continuous superfusion. A variety of perfusion chambers with either open or closed configurations are available.

Sometimes it is necessary to design and construct very sophisticated temperature and fluidlevel control systems. Such chambers permit very rapid exchange of medium, which is necessary for physiological experiments requiring high time resolution. There are also now commercially available, programmable, automated perfusion systems that permit rapid switching between one of several perfusion channels. Some experimental conditions require only relatively simple, low-cost chambers and perfusion systems.

The tissues seem to remain healthy for at least 20 hr when perfused (10-20 ml/hr) with either the culture medium or normal saline, both of which are buffered with 25 mM HEPES. Specimen heating is essential for many experiments, but this can induce an agonizing battle with focus stability as the chamber and stage components heat up. Because there is always a time lag between when the sensor of the temperature control detects that the temperature is too high (or low) and the time that the heater is able to warm the whole stage, the actual temperature of most stage heater is always slowly oscillating, a fact that causes the focus plane to shift in a periodic manner.

A sufficient period of preheating can sometimes reduce this problem. Another approach is to use a modified hair dryer to blow warm air onto both the chamber and the stage or an egg-incubator heater to heat all the air in an insulated box surrounding the entire microscope. It is also important to monitor the temperature of the perfusing medium very near to the specimen. A low-cost microprocessor temperature controller that reduces fluctuations in the heating/cooling cycle can be obtained from Omega Engineering.

Imaging methods

It should by now be evident that a primary concern when imaging living cells is photon collection efficiency. This is especially true when imaging dynamic processes, such as cell migration, over long periods of time. Higher collection efficiency will afford effectively lower excitation light levels, thus permitting more frequent sampling or observations of longer duration. Commercially available laser scanning confocal systems. For illumination, the microscopes arc equipped with multiple lasers. Both systems offer simultaneous excitation and detection in more than one epi-fluorescent channel. In addition, the SP2 AOBS system offers increased flexibility by providing continuously variable wavelength selection in the emission pathway. This permits customizing the spectral detection to maximize throughput in separate channels.

Imaging deep within tissue

Often the goal of studies in tissue slices is to examine biological processes within a cellular environment that approximates that found in situ. In the case of tissue slices, it is usually desirable to image as far from cut tissue surfaces as possible to avoid artifacts associated with tissue damage. However, the cut surfaces of developing CNS tissue slice cultures contain a plethora of astrocytes, activated microglia, and a mat of growing neuronal processes. Time-lapse imaging of these regions provides striking footage of glial cell movements, proliferation, and phagocytosis. With oil-immersion lenses, useful fluorescence images seem to be limited to a depth of 50-75 μm or so into the tissue. Imaging deeper (> 50 μm) within tissue can suffer from several factors, including:

— Weak staining of cells due to poor dye penetration

— Light scatter by the overlying tissue components

— Spherical aberration

The first problem can be overcome if the dye can be injected into the tissue with a minimum of disruption, or if tissues can be harvested from transgenic animals expressing fluorescent proteins in subsets of cells. Light scatter by the tissue can be minimized by using longer-wavelength dyes. Indeed, find that a long-wavelength conjugate of IB4 noticeably improves visibility of labeled cells in deeper portions of the tissue slice. Imaging at longer wavelengths may also reduce phototoxic effects since the light is of lower energy. Finally, the problem of spherical aberration can be improved by using water-immersion objective lenses or the recently introduced automatic spherical aberration corrector.

Keeping cells in focus

The optical sectioning capability of the confocal microscope can be simultaneously a blessing and a curse. On the one hand, thin optical sections reduce out-of-focus flare and improve resolution. However, with such a shallow depth of focus, even very small changes in the position of the objective lens relative to the object of interest within the specimen can ruin an otherwise perfect experiment. This is a particular problem when imaging thin, tortuous structures such as axons or dendritic spines within neural tissue.

A moving focal plane can, for example, give one the erroneous impression of dendritic spine extension or retraction. This problem is compounded when imaging cells and cell processes that are in fact actively moving within the tissue. One obvious approach is to image the cells in four dimensions (3D x time). This can absorb some changes in tissue and stage movement as well as help track cells that move from one focal plane to another. In addition, our strategy has been to image with the detector pinhole aperture substantially open, corresponding to a pinhole size roughly 4 Airy units in size.

Although this reduces the axial resolution slightly, it has the dual advantage of achieving a higher signal-to-noise ratio at a given illumination intensity as well as thickening the optical section. The open pinhole configuration gives an apparent optical section thickness of about 3 μm when using a 20x NA 0.7 objective. Thus, for each time point, here collect about 15 images at axial step intervals of ~2 μm. The guiding principle here is to space the image planes in the axial dimension so as to maximize the volume of tissue imaged but not lose continuity between individual optical-section images.

When these image stacks are collected at ~5-min intervals at power levels of ~50-75 μW, IB4-labeled cells do not appear to suffer phototoxic effects and can be imaged continuously for over 20 hr. Image stacks can be recombined later using a maximum brightness operation. Unfortunately, even when z-axis stacks of images are collected, tissue movements can be so severe as to necessitate a continuous "tweaking" of the focus.

Thus, it is helpful to monitor image features on the screen, or to store the data in such a way that they are quickly accessible and can be reviewed on the fly to make corrective focus adjustments. Ideally, one would like an automated means of maintaining the desired plane of focus, especially for long imaging sessions. Although there are several autofocus methods that work for simple specimens, imaging structures in 3D tissue presents a significant challenge because there is no single image plane on which to calculate focus.

Handling the data

Imaging tissue in 3D over time solves some problems but generates others. Fortunately, improvements in desktop computer performance and storage capacity, coupled with low cost, make this much less of a problem than it used to be. Desktop workstations now contain hard drives with very large storage capacities, and archiving methods and media are readily accessible and inexpensive. Typically store newly acquired data on a network server for image processing, and analysis, then archive the data onto CDs or DVDs.

This is a technology that is sure to continue rapid advancement, making it easier to store and access image data. At the end of the time-lapse experiment, the z-axis image stacks are combined in a variety of ways for viewing. For time-lapse studies, it is generally most useful to produce a set of "extended focus" images for viewing time-points in rapid succession. Depending on the file format, these image series can be viewed in one of a number of image viewers, including Scion Image or ImageJ.

To create the projection images, use a maximum brightness operation running in a custom-written Pascal macro to construct a 2D representation of the 3D data set. When the axial step interval is appropriate, portions of single cells that pass through the various focal planes appear contiguous. Alternatively, the image stacks can be reassembled into a set of red-green 3D stereo images. Such images, when played in rapid succession, provide 3D-depth information as well as time information in thick tissue samples. If more than one fluorescent channel is used, they can be combined to create multicolor images.

CHAPTER 9

MICROBIAL CELL FACTORIES

Production of properly folded proteins with high yield and purity may not always be achieved. Issues such as folding, solubility, protein stability, transcription and translation efficiency, posttranslational processing, secretion, metabolic burden and other stress responses resulting from recombinant protein production, as well as protein purification, need to be addressed in order to obtain biologically active recombinant proteins with high purity and yield. In this regard, genetically encoded fluorescent reporters provide ample new opportunities to better tackle these issues.

Since the demonstration of the Aequorea victoria green fluorescent protein (GFP) as a versatile reporter, several additional GFP-like fluorescent proteins with various colors have been discovered and their genes cloned. Synthetic fluorescent protein variants have also been developed, exhibiting traits distinct from their wild-type counterparts. The properties of selected fluorescent protein variants derived from the A. victoria GFP and the Discosoma red fluorescent protein (DsRed) are summarized in Table 1.

Fluorescence spectra of enhanced GFP variants along with DsRed are shown in Figure 1, and fluorescence of purified protein variants derived from DsRed are shown in Figure 2. These GFP-like proteins each has its own unique properties, while sharing common structural, biochemical and photophysical characteristics.

Table 1: Properties of selected fluorescent proteins

Fluorescent Protein	Excitation Peak (nm)	Emission Peak (nm)	Extinction Coefficient (M^{-1} cm-1)	Fluorescence Quantum Yield
EBFP	383	445	31,000	0.25
ECFP	434	477	26,000	0.40
Cerulean CFP	433	475	43,000	0.62
EGFP	489	508	55,000	0.60
EYFP	514	527	84,000	0.61
Venus YFP	515	528	92,200	0.57
Citrine YFP	516	529	77,000	0.76
DsRed	558	583	75,000	0.79
mRFP1	584	607	50,000	0.25
mHoneydew	487/504	537/562	17,000	0.12
mBanana	540	553	6,000	0.70
mOrange	548	562	71,000	0.69
mTangerine	568	585	38,000	0.30
mStrawberry	574	596	90,000	0.29
mCherry	587	610	72,000	0.22

GFPlike proteins are relatively small (25-30 kDa) and their fluorescence mechanism is self-contained, requiring no cofactors. These unique properties make GFP-like proteins very attractive tools in non-invasive biological monitoring applications. As a tool to improve recombinant protein production, fluorescent proteins can be used to monitor the protein product or the cellular processes relevant to recombinant protein production.

Monitoring Protein Production

Fluorescent proteins are commonly used as a reporter for a protein of interest, normally by tagging the fluorescent protein reporter to the protein of interest via genetic fusion. Functional fusion of Aequorea GFP to a broad range of protein partners at either N- or C- terminus has been reported, and a direct quantitative correlation between the GFP fluorescence intensity and the titer or even the functional activity of the fusion partner can often be established.

To minimize potential interference by the GFP tag on its fusion partner, it is desirable and sometimes necessary to incorporate a peptide linker to allow sufficient spatial separation of the two protein moieties to assure fusion protein stability and functionality. Flexible linkers lacking large bulky hydrophobic residues (e.g. GSAGSAAGSGEF) are commonly used, while hydrophilic helix-forming linker peptides have been reported to be superior to flexible linkers in some cases. To allow removal of the GFP tag, an enzymatic cleavage site (e.g. enterokinase or Factor Xa cleavage sites) can be engineered into the linker.

It is preferred to splice the GFP/linker to the N-terminus of the target protein, provided such fusion does not impair the target protein function and stability. With the majority of the enzymes commonly used for tag removal, this fusion orientation enables elimination of the tag without leaving extraneous amino acid residues on the target protein after cleavage. Alternatively, chemical cleavage based on cyanogen bromide, formic acid, or hydroxylamine may be considered, provided the target proteins are not susceptible to cutting by these chemical agents.

Further information of tag removal can be found in a comprehensive review by Hearn and Acosta. In addition to tandem fusion, insertional fusion (i.e. by inserting the protein of interest into GFP or vise versa) may also be feasible. Recombinant protein production can be monitored non-invasively, in situ, and almost in real time, by monitoring culture fluorescence using on-line optical sensors. This information is useful in determining the optimal product harvest time to avoid product degradation and to devise process control strategies to optimize culture/ operating conditions to improve recombinant protein production.

GFP has also been used to monitor recombinant virus titers in cell cultures, and cell density in microbial, animal, and plant cell cultures; the cell growth information can be used, in turn, to optimize the culture process for improved recombinant protein production (e.g. by optimizing the feeding profiles of the limiting nutrient or the promoter inducer, or by determining the optimal product harvest time). Additionally, GFP-fusion coupled with flow cytometric analysis is useful for profiling recombinant protein expression among different cell subpopulations, and selection of high-producing cells.

GFP-fusion can also be used for monitoring protein secretion and other subcellular protein localization and trafficking events. In the event direct GFP fusion hampers protein secretion, alternative protein fusion strategies may be sought. One plausible approach is to express GFP and a protein of interest as a cleavable chimeric polyprotein. By targeting the polyprotein to the secretory pathway, the target protein and the GFP tag may become separated and secreted as individual proteins. Feasibility of such approach has been demonstrated in fungi and plant cells for the successful in-vivo cleavage of an glucoamylaseinterleukin-6 fusion protein and a fusion antimicrobial polyprotein, respectively.

Cellular processing in plant cells of a polyprotein that consists of DsRed-GFP fusion linked by a Kex2 cleavage sequence is currently being investigated in the author's laboratory. In another possible approach, one may express the target gene and GFP in a dicistronic vector by incorporating an internal ribosome entry site (IRES). One additional possi.bility would be to express target protein and GFP as separate genes, but from the identical promoter. If a constant ratio between GFP fluorescence and target protein concentration could be established, the independent GFP could be used for fluorescent monitoring. Fluorescent protein reporters can also be used to

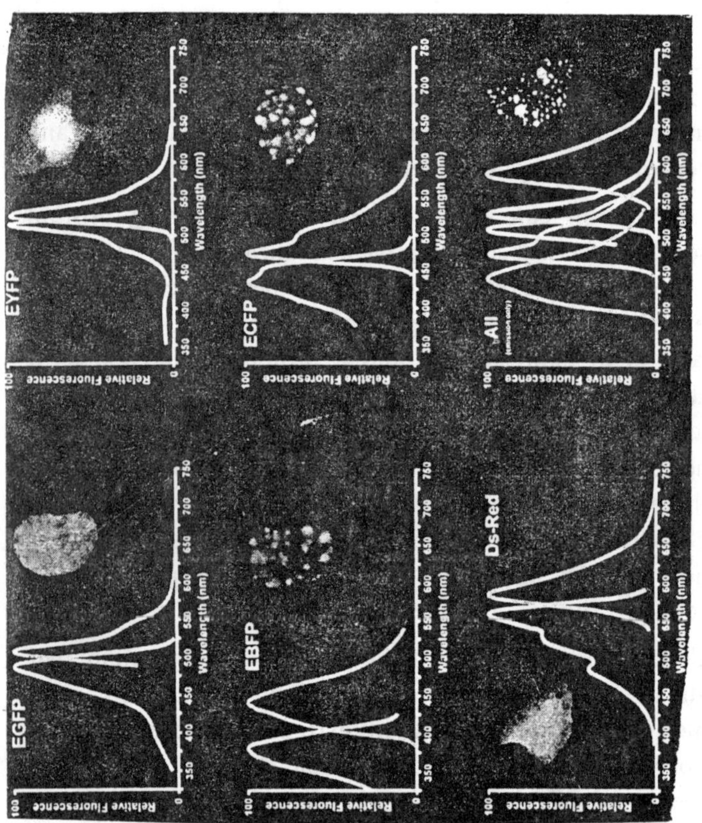

Figure 1. Fluorescence spectra of EGFP variants and DsRed.

probe protein-protein interactions that regulate the protein secretion process, potentially leading to development of molecular strategies that improve recombinant protein secretion.

Monitoring Protein Purification

The fact that GFP fluorescence is readily detectable makes it a very attractive tool for optimizing purification of recombinant proteins. Poppenborg et al. optimized immobilized metal affinity separation of a histidine-rich protein tagged with GFP by tracking the fluorescence of the fusion protein. Since GFP is a highly hydrophobic protein, recovery of GFP-fusion proteins can be facilitated by using hydrophobic interaction chromatography (HIC).

Figure 2. Fluorescence of purified protein variants derived from DsRed

GFP has also been engineered to allow affinity purification. Paramban et al developed a chimeric GFP tag having an internal hexa-histidine sequence. Such a GFP tag allows efficient purification of GFP-fusion proteins based on immobilized metal affinity separation, as well as maximum flexibility for protein or peptide fusions since both termini of the GFP are available.

Monitoring Protein Folding

Correct folding is one of the key challenges in recombinant protein production using simple hosts like Escherichia coli. GFP has been proposed as a folding reporter. By fusing GFP to a panel of proteins, Waldo et al demonstrated that display of GFP fluorescence in the E. coli colonies expressing the fusion proteins indicated proper folding of GFP's fusion partner. De Marco however cautioned that observation of GFP fluorescence from the fusion protein may not guarantee the fusion partner have reached its native structure.

Recently, Waldo and coworkers reported a novel split-GFP system that consists of a small (GFP β strand 11; GFP 11) and a complementary large fragment (GFP β strand 1-10; GFP 1-10). Neither fragment by itself displays fluorescence, but GFP fluorescence emerges upon self-association of the two complementary fragments. These researchers demonstrated that by tagging target proteins with the GFP-11 tag, the solubility of the target proteins can be checked by mixing with the GFP 1-10 fragment in vitro or by co-expressing the GFP 1-10 fragment in vivo. Appearance of GFP fluorescence suggests the target protein is soluble. This split-GFP may also be useful for detecting protein-protein interaction in vivo, similar to the luciferase fragment complementation systems.

Monitoring Cellular

In addition to direct monitoring of protein expression, processing, and secretion, fluorescent proteins can also be used to monitor cellular events that are directly or indirectly related to recombinant protein production. For instance, FRET (fluorescence resonance energy transfer)-based GFP sensors have been developed to detect proteolysis in vivo. FRET-based GFP sensors have also been developed for measuring intracellular concentration of nitric oxide, calcium, cAMP, zinc, activation of G protein-coupled receptor (GPCR), and PKA-mediated phosphorylation.

Redox sensitive GFP variants have been developed for monitoring the cellular oxidative states, which strongly affect protein folding. In addition, intracellular pH can be measured using GFP. Metabolic stresses induced by recombinant protein over-expression may also be monitored in vivo using GFP linked to stressinduced promoters as a reporter.

OPTICAL SENSING

The presence of cell aggregates, debris, and other light absorbing/scattering compounds in the culture medium contributes to the "inner filter effect" (IFE) that could distort the optical measurement of culture GFP fluorescence. Real-time compensation of IFE in monitoring cultures expressing GFP-fusion proteins typically involves establishing a mathematical model to link the IFE to cell density, and to use an on-line laser turbidity sensor to report the biomass density needed in the calculation of the IFE.

An obvious drawback of such an approach is the requirement of a turbidity sensor in addition to the optical sensor for monitoring culture fluorescence. A technique that allows real-time compensation of IFE during on-line monitoring of culture GFP fluorescence, without the need for an additional biomass sensor. This was achieved by developing a model-based state observer, using the extended Kalman filter (EKF) and on-line measurement of GFP culture fluorescence using an optical light-rod sensor.

Multiple Fluorescent Protein

Given the many facets of its applications, multiple fluorescent protein reporters could potentially be used in parallel for multi-color in vivo sensing. For instance, GFP may be used to tag the recombinant protein product, while a red fluorescent protein could be linked to a stress-responsive promoter such as the heat-shock promoter groEL to monitor stress induced by recombinant protein overexpression in E. coli. Having a large repertoire of fluorescent proteins with diverse spectral properties is also necessary for multiplex FRET-based sensing applications.

Through both structure-based modification and evolutionary methods for protein engineering, several robust variants of the Aequorea GFP have been created with blue, cyan, and yellow colors. In a recent paper by Shaner et al, development of a panel of novel monomeric red, orange and yellow fluorescent proteins derived from the Discosoma DsRed was reported. These new variants also show additional traits that are useful for monitoring recombinant protein production.

Monomeric Fluorescent Protein

Wild type DsRed is an obligate tetramer. When fused to a protein of interest, the fusion protein often forms aggregates, hampering normal localization, trafficking and protein-protein interactions of the protein of interest. A monomeric DsRed variant, called mRFP1, was developed by disrupting each subunit interface via insertion of arginines, and then using directed evolution to accelerate chromophore maturation and to restore fluorescence, which takes 33 substitutions. mRFP1 was further improved by subjecting to additional rounds of directed evolution.

The resulting eight variants display corresponding emission peaks ranging from 537 to 610 nm. Some of these new variants also show better tolerance to N- and C- terminal fusions, higher extinction coefficients, quantum yields, and photostability, though no single variant has acquired all the desirable traits. Since GFP is known to have high tolerance to either N- or C- terminal fusions, and DsRed and GFP share similar structures, Shaner et al engineered GFP-type termini into mRFP1, rendering improved tolerance to protein fusion in the new variant.

Among the new monomeric variants, mCherry (excitation at 587 nm, emission at 610 nm) is the most red-shifted, and has the best photostability, fastest maturation (15 min) and excellent pH resistance and tolerance to N-terminal fusions. The mOrange variant (excitation at 548 nm, emission at 562 nm) has high extinction coefficient and quantum yield, and is shown to be a superior FRET acceptor for GFP variants.

CHAPTER 10

PRODUCTS OF INTEREST OF INDUSTRY

PHARMACEUTICALS

Alkaloids

Different types of alkaloids have been used as pharmaceuticals and most of them are plant metabolites. The typical tropane alkaloids, atropine, hyoscyamine, scopolamine and cocaine, were widely used as blockers of the parasympathetic nervous system such as anodyne and antispasmodic. Research on production of these useful alkaloids by plant cell cultures has been carried out for more than 25 years, however, industrial production has not yet succeeded because of low producing ability of the cultured cells. Plants used for these studies are mainly Atropa belladonna, Hyoscyamus niger, Datura meteloids and others.

Morphinan alkaloids

Codeine is an analgesic and cough-suppressing drug and Papaver somniferum L. (opium poppy) is a traditional commercial source of codeine, and morphine which can be converted to codeine. Mature capsules of P. bracteatum accumulates up to 3.5% of thebaine which also can be converted to codeine. Although many researchers have tried to produce codeine by undifferentiated cells of these plants, little success has been achieved.

Concerning sanguinarine, Eilert et al. showed that fungal mycelium of Botrytis sp. elicited production of sanguinarine by P. somniferum cells. The level of this alkaloid increased 26 times in the presence of the elicitor, 29% of the dry cell weight, compared to the medium without it. Since the alkaloid extracted from intact plants is added to toothpastes as an anti-plaque agent, the commercial production of sanguinarine by cell culture technology was intensively investigated by Kurz's group at the National Research Council of Canada and a company in the U.S. several years ago, but the process has not yet been commercialized. The efficient production of thebaine and codeine using cell culture systems by de novo synthesis was not successful, Furuya et al., therefore studied the biotransformation of codeinone to codeine using immobilized cells of P. somniferum. The conversion yield was 70.4% and about 88% of codeine converted was excreted into the medium.

Berberine

Berberine is an isoquinoline alkaloid which is distributed in roots of Coptis japonica and cortex of Phellondendron amurense. Berberine chloride is used for intestinal disorders in the Orient and it takes 5 to 6 years to produce Coptis roots as the raw material. Furuya at Kitasato University and Yamamoto at Nippon Paint have investigated the production of berberinc by Coptis japonica cell cultures since 1970's and Yamada et al. at Kyoto University selected a high berberine producing cell line of C. japonica which was transferred to a Japanese company, Mitsui Petrochemical.

Mitsui Petrochemical has improved the productivity, and Hara et al. found that addition of 10-8 M gibberellic acid into the medium stimulated berberine productivity up to 1.66 g per L of the medium. Using a cell sorter and protoplasts of C. japonica, they selected many higher alkaloid-producing cell lines. Thus, the Mitsui group produces berberine in a large scale at a level of 1.4 g per L of their optimized medium within 2 weeks. Furthermore, they established a "high-density cell culture" process to produce berberine much more efficiently. In order to achieve a cell mass of 70 g/L on a dry weight basis, stirring without damaging the cells, supply of sufficient amounts of oxygen, and that of appropriate nutrients were optimize.

As a result, the yields reached 70 g per L of cell mass and 0.45 g/day of berberine. The continuous culture with high cell density was also conducted successfully. Addition of a polyamine, spermidine, was found to stimulate the production of berberine by Thalictrum minus cell suspension cultures by Hara et al. in Tabata's laboratory although other polyamines such as cadaverine, putrescine and spermine were ineffective. They indicated that spermidine effected an increase of ethylene generation which was associated with activation of berberine synthesis. The maximum stimulative effect was obtained by addition of 2 mM spermidine.

Tropane alkaloids

Scopolamine and hyoscyamine are being used commercially as anesthetic and antispasmodic drugs. These alkaloids occur in leaves of (Solanaceae) plants including D. myoporoides and D. leichhardtii. Scopolia, Atropa, Hyoscyamus and Datura also contain tropane alkaloids. Studies on production of tropane alkaloids by plant tissue cultures have been actively carried out by many researchers since West et al. found tropane alkaloids in an Atropa belladonna callus more than 30 years ago. However, the concentrations of scopolamine and hyoscyamine in cultured cells are generally very low in spite of many efforts to increase the yield using various approaches.

Therefore, the plant cell culture has not yet been employed to manufacture these alkaloids. For example, Tabata et al. added tropic acid into Scopolia japonica suspension cultures as a precursor and could increase the level of alkaloids up to 15 times. Mitsuno et al. selected a high tropane alkaloid producing strain of Hyoscyamus niger which produced about 7 times more hyoscyamine (13.9X10-3% fresh cells), than that of the parent strain. According to their results, there was no direct correlation between high producing ability and variation of chromosomal numbers.

Endo et al., Kitamura et al. and many other scientists reported that roots differentiated from cultured cells accumulate scopolamine, hyoscyamine and/or nicotine. However, Kitamura indicated that the alkaloids were not accumulated in leaves of the regenerated plantlets. Since the alkaloids are synthesized in the root, Flores et al. cultivated hairy roots transformed with Agrobacterium rhizogenes and showed the production of hyoscyamine and other alkaloids at the similar levels to the normal roots.

Cardinolides

Cardiac glycosides or cardenolides are products of Digitalis species. Some of these compounds have been employed in treatment of heart diseases. Chemical structures of the major aglycone of cardenolides are shown in Fig. 1 and various sugar residues link the 3-hydroxy group of aglycones. For commercial production Digitalis plants are being cultivated in the fields of several countries including the Netherlands, Hungary and Argentina. A digoxin product, Lanoxin, is a brand name of a Burroughs Wellcome product and has the largest market of the company's cardiovascular drugs. The major markets of Lanoxin are in the U.S.A. and Italy, and the total sales are approximately 6000 kg a year with 50 million U.S. dollars. Other companies such as Boehringer Mannheim, Merck Darmstadt and Beiersdorf AG in Germany also sell cardiac glycosides.

The studies of plant tissue and cell cultures for production of cardiac glycosides were begun more than 30 years ago and the report presented by Hildebrandt and Riker in 1959 was probably the first one in this field. Staba investigated the nutritional requirements of tissue cultures of Digitalis lanata and D. purpurea in 1962. These two species are being commonly used by many scientists. Although there are a number of papers describing production of cardiac glycosides in Digitalis tissue cultures, generally the yield was very low, and moreover, during the successive transfers of the cultured cells the amount of cardenolides often decreased and disappeared completely.

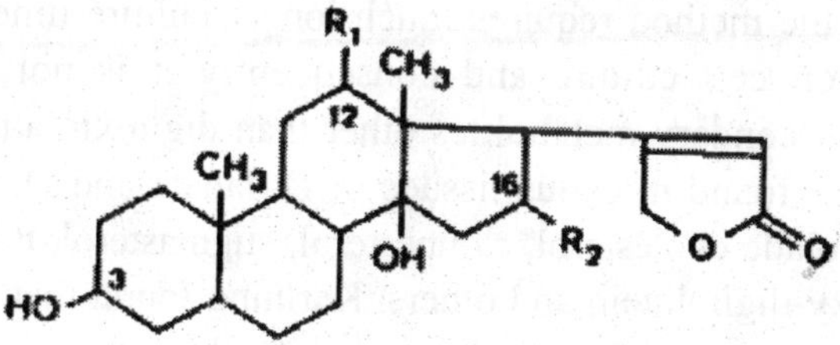

	R_1	R_2
DIGITOXIGENIN	H	H
GITOXIGENIN	H	OH
GITALOXIGENIN	H	OCHO
DIGOXIGENIN	OH	H
DIGINATIGENIN	OH	OH

Figure 1: Principal Cardioactive Glycosides of Digitalis species: Chemical Structures

Instead, many researchers indicated that morphological differentiation caused an increase in productivity. For example, Lui and Staba showed that organ cultures of D. lanata leaves and roots produced cardenolides and during the cultivation the level of digoxin in the tissues rose with increasing age. Hirotani and Furuya also found that renewed organ differentiation from callus tissues of D. purpurea led to a new formation of cardinolides. Nutritional factors including growth regulators, sugars, nitrogen sources, vitamins and so on in the medium affect on differentiation of shoots and other organs and the secondary metabolites such as cardiac glycosides are often synthesized.

Hagimori et al. of Japan Tobacco Inc.cultivated shoot-forming tissues of D. purpurea in a 3 L jar fermentor and detected high concentrations of cardiac glycosides including digitoxin. The differentiated tissue culture is prerequisite for secondary metabolites production in some cases, however in general, the method requires much longer culture time than the suspension cell culture and consequently it is not efficient. Various secondary metabolites other than digitoxin and digoxin have been found in callus tissues of D. lanata and D. purpurea. These include cholesterol, campesterol, stigmasterol, ß-sitosterol, 4-hydroxy-digitolutein and others. Kartning found that the levels of those secondary products tended to decrease over several passages in cultivation.

In contrast to the de novo synthesis, the biotransformation process with Digitalis plant cells seems to be more promising from a commercial point of view. Graves and Smith reported that D. lanata and D. purpurea callus cultures rapidly transformed progesterone to pregnane. Leaf and root cultures of D. lanata and shoot-forming callus tissues of D. purpurea accumulated an increased level of digoxin and/or digitoxin when progesterone was added to the cultures. Stohs and Staba studies on the biotransformation of cardenolides by Digitalis cells and recognized that the glycosylation reaction occurred.

Among many studies, the biotransformation from digitoxin to digoxin using Digitalis lanata cells investigated by Reinhard and Alfermann is the most interesting approach in terms of commercial application since digoxin has a higher demand as a drug for heart diseases than digitoxin. It is advantageous that Digitalis leaves contain a larger amount of digitoxin which can be used as a substrate. It is a hydroxylation reaction at the 12 ß-position of digitoxin, and Reinhard et al. found that ß-methyldigitoxin was the most suitable substrate in this biotransformation as methyldigoxin is the major product. To reduce the production cost, the same group examine the use of immobilized D. lanata cells and semi-continuous culture.

L-DOPA

L-DOPA, L-3,4-dihydroxyphenylalanine, is an important intermediate of secondary metabolism in higher plants and is known as an precursor of alkaloids, betalain, melanine, and others. It is also a precursor of cathecolamines in animals and is being used as a potent drug for Parkinson's disease. Brain in 1976 found that the callus tissue of Mucana pruriens accumulated 25 mg/L DOPA in the medium containing very high concentration of 2,4-D. Wichers et al. immobilized cells of M. pruriens within alginate and found that the cells produced DOPA from tyrosine in up to 2% of dry cell weight. The DOPA synthesized was secreted mostly into the medium.

Teramoto and Komamine induced callus tissues of Stizolobium hassjoo (Mucuna hassjoo), M. pruriens and M. deeringiana, and optimized the culture conditions. The highest concentration of DOPA was obtained when S. hassjoo cells were cultivated in MS medium containing 0.025 mg/l 2,4-D and 10 mg/l kinetin. The level of DOPA in the cells was about 80 nmol/g-f.w.

Valepotriates

Plants in the Valerianaceae have been used as folk medicines. For example, Nardostachys jatamansi, Valeriana wallichii and V. officinalis L. var. angustifolia have been used in India, and N. chinensis has been employed in China for hundreds of years. Partrinia plants are also being used as sedative drugs in former Soviet Union. Although the active principles in these plants have not been identified, a group of compounds having biological activities such as sedative, tranquilization, cytotoxicity and antitumor activities were named "valepotriates". Thies synthesized a series of valepotriate derivatives and tested for biological activities.

Becker et al. have investigated plant tissue cultures for producing valepotriates because of their limited supply and uncertain availability. They induced callus tissues of nine different species of Valerianaceae on MS media and found that Fedia cornucopiae and V. locusta cells produced higher levels of the compounds than the intact plants. Isolation of cell lines resistant to trifluoroleucine and to nystatin, treatment with colchicine, cultivation of the cells in two-phase culture media and addition of several bioregulators were carried out intensively. As valepotriates have monoterpene skeletons, L-leucine was considered as a precursor.

Baker et al. isolated cell lines of V. wallichii resistant to trifluoroleucine, a leucine analog, but the yield of valepotriates in the cells was not increased although the intracellular level of leucine was increased. It was reported that fungi resistant to nystatin, a polyene antibiotic, produced a high level of steroids. One of the strains of V. wallichii to resistant to nystatin increased the amount of valepotriates produced up to 3 times, which was 88 mg/g-d.w. The same group treated a suspension culture of V. wallichii with colchicine which was expected to induce polyploid cells. As a result, they found that the colchicine treated cells produced higher amount valepotriates than the respective untreated cultures.

A two-phase culture provided by addition of RP-8(Merck) into the culture medium was effective in inducing secretion of the lipophilic compounds from the cells, and the total yield of valepotriates was substantially increased. Since the valepotriate skeleton is of the iridoid nature, they added some plant bioregulators such as dimethyl-morpholinium-bromide, dimethyl-piperidinium-bromide, dimethyl-piperidinium-chloride as well as 2-(3,4-dichloro-phenoxy)-triethylamine and 2-(3,5-

diisopropylphenoxy)-triethylamine to cell suspension cultures of F. cornucopiae and V. wallichii. When they were employed in concentrations of 0.01 to 0.04 mmol during the early exponential growth stages of the cells, the levels of valepotriates were increased significantly.

Antitumor Compounds

The plant kingdom is one of the attractive sources of novel antitumor compounds. The National Cancer Institute in the U.S., for example, has conducted an intensive screening program since 1955 and has identified various potent compounds from higher plants. These antitumor compounds include maytansine, tripdiolide, homoharringtonine, bruceantin, ellipticine, thalicarpine, indicine-N-oxide, and baccharin. In addition to these compounds, some of the higher plant products such as vinblastine, vincristine, podophyllotoxin derivatives including etoposide, and camptothecin and its derivative have already been marketed as very important anticancer drugs. Taxol from Taxus brevifolia and related plants, is one of the most exiting compounds and was marketed in 1992.

However, the concentrations of these active compounds in plants are generally low, the growth rate of the plants is slow and the accumulation pattern of these compounds is highly susceptible to geographical or environmental conditions. Therefore, it is not an easy task to produce economically these compounds by extraction from intact plants. Furthermore, as indicated for taxol production, the exhaustion of native producing-plants is becoming a serious problem in terms of environmental preservation.

Though plant tissue culture processes are still not cost-effective if targeted products could easily be manufactured by chemical, fermentative and/or extraction processes, this technology is undoubtedly one of the appropriate approaches to solving the above-described problems. And many interesting antitumor compounds isolated from higher plants have very complicated chemical structures. Therefore, the research interest has grown for over a decade and a couple of processes based on the plant tissue culture technology are likely to be applied for commercial production of antitumor drugs. The following are examples of the technology.

Camptothecin

Camptotheca acuminata, a native of North China, was found to produce a potent antitumor alkaloid, camptothecin, by Wall et al. in 1966. It is highly active in Walker 256 rat carcinosarcoma and mouse leukemia, p388 and L1210. The clinical trials in patients with gastrointestinal cancer were at first very promising but subsequent trials showed toxicity. Sakato and Misawa induced callus from the stem of C. acuminata on MS solid medium containing 0.2 mg/L 2,4-D and 1 mg/L kinetin.

The callus was transferred to the liquid medium. Gibberelin, L-tryptophan and conditioned medium stimulated growth of the cells. After 15 days of cultivation in suspension, the concentration of camptothecin in the cells was 0.0025% on a dry weight basis, which was about 1/20 of the level in the intact plant. A.J. van Hengel et al. in the Netherlands established the suspension culture system of C. acuminata and detected camptothecin in the cultured cells using TLC, HPLC and GC-MS. The highest level, 0.998 mg of camptothecin per liter of the medium, was accumulated in the cells cultivated in MS medium containing 4 mg/L NAA. 10-Hydroxycamptothecin having less toxicity is so far a promising derivative of camptothecin and is in clinical trials in the U.S.

Homoharringtonine

Homoharringtonine together with harringtonine and isoharringtonine, were isolated from Cephalotaxus harringtonia by Powell et al. in 1969. These alkaloids are complex esters of the inactive alcohol, cephalotaxine. These compounds inhibit the growth of murine leukemias, L1210 and P388, and KB cells. Homoharringtonine also shows activity against colon tumors, melanoma and leukemia in mice. Since the chemical synthesis of homoharringtonine was not efficient for commercialization, studies of tissue culture were conducted by Delfel and his colleagues (187). They detected about 5-10 mg of the alkaloids per kg dry wt. of the callus tissues cultivated for 3 to 6 months.

The levels were approximately 1 to 3% of the concentrations found in the parent plant. These products were cephalotaxin, homoharringtonine, harringtonine and isoharringtonine. MS medium containing 1 mg/L kinetin and 3 mg/L NAA was favorable to induce the callus from C. harringtonia, while the medium without growth regulators strongly promoted organogenesis. The radioimmunoassay established by the same group showed the levels of cephalotaxine and its esters were only 1/300 of the intact plant.

Podophyllotoxin

Podophyllum pelatatum, May apple, which is a common herb in eastern North America contains an antitumor lignan, podophyllotoxin. It is active to KB cells and is used against certain virus diseases and skin cancer. A semi-synthetic derivative of podophyllotoxin, etoposide (V-16), was found to be active against brain tumor, lymphosarcoma and Hodgkins' disease and was approved by the FDA in the U.S. Bristol-Myers Squibb is one of the largest manufacturers of the drug.

Production of podophyllotoxin by P. pelatum cell cultures was first attempted by Kadkade and he found that a combination of 2,4-D and kinetin in the medium supported the highest amount of its production. Red light stimulated the production. Sakata et al. of Nippon Oil induced embryogenic roots from a callus of the plant in a liquid MS medium supplemented 1 mg/L NAA, 0.2 mg/L kinetin and 500 mg/L casein hydrolysates. The roots were then transferred to the medium without growth regulators. They detected 1.6% of podophyllotoxin in the dried tissues, which was 6 times higher level than that in a mother plant.

To increase the yield of podophyllotoxin, Woerdenberg et al. in the Netherlands added a complex of a precursor, coniferyl alcohol, and ß-cyclodextrin to Podophyllum hexandrum cell suspension cultures. Addition of 3 mM coniferyl alcohol complex gave 0.013% podophyllotoxin of the cells on a dry weight basis but the cultures without the precursor produced only 0.003%. ß-D-glucoside of coniferyl alcohol, coniferin, was a more potent precursor in terms of the yield of the anticancer compound (0.055%), but unfortunately this compound is not commercially available. The same authors reported that cell suspension cultures of Callitris drummondii (conifer) also accumulated podophyllotoxin-ß-D-glucose. In the dark, the cells produced approximately 0.02% podophyllotoxin of the dry cell mass and 85-90% of the lignans were the ß-D-glucoside form, while in the light the yield of podophyllotoxin-ß-D-glucose increased to 0.11%.

Smooly et al. reported that callus tissues and suspension culture cells of Lilium album produced podophyllotoxin. One of the cell lines produced 0.3% podophyllotoxin of dried cells together with small amounts of 5-methylpodophyllotoxin, lariciresinol and pinoresinol after 3 weeks of cultivation. The callus tissue induced from P. hexandrum was reported by Heyenga to produce podophyllotoxin, 4'-demethyl-podophyllotoxin and podophyllotoxin-4-0-glucoside when the callus was incubated in B5 medium containing 2,4-D, gibberellic acid and 6-benzylaminopurine. The levels of podophyllotoxin and its derivatives were similar to those in the mother plant.

Vinca alkaloids

The dimeric indole alkaloids, vinblastine and vincristine have become highly valued drugs in cancer chemotherapy due to their potent antitumor activity against various leukemias, Hodgkin's disease and solid tumors. They are currently produced commercially by extraction from Catharanthus roseus (Apocyanaceae) plants, but the process is not efficient because of very low concentrations of the alkaloids in the plant. It was reported that the concentration of both vinblastine and vincristine was only 0.0005% as a dry weight basis.

In order to produce these useful anticancer drugs much more efficiently, many scientists have tried to apply plant tissue culture technology. The concentration of vindoline in the intact C. roseus plant is approximately 0.2% as a dry weight basis, which is much a higher level than catharanthine, and the cost of vindoline is less expensive compared to catharanthine and vinblastine. The Allelix group, therefore, investigated the production of catharanthine by a cell suspension culture process with a selected C. roseus cell line induced from anthers on Gamborg's B5 medium containing 2% sucrose, 1 mg/L 2,4-D and 0.1 mg/L kinetin. The cells were grown in 250 ml flasks containing 60 ml of MS liquid medium supplemented with 3% sucrose, 1 mg/L NAA and 0.1 mg/L kinetin under continuous diffuse light on a rotary shaker (250 r.p.m.) at 25° C.

In experiments for optimization of catharanthine production, they transferred 7 day old cells to a test medium and subcultured for 3 passages. In the 4th passage, 60 ml cultures were harvested in triplicate after 2 or 3 weeks growth, and the cell mass and alkaloid content were determined.

The results showed that the MS medium was the most favourable for catharanthine production but the optimal levels of phytohormones for the growth and the production were varied in different cell lines. For example, one line required no phytohormones but another line required 0.1 mg/L NAA and 0.1 mg/L kinetin. Addition of various chemically defined compounds to the medium as "inducers" was found to stimulate the production efficiently. Among them effects of vanadyl

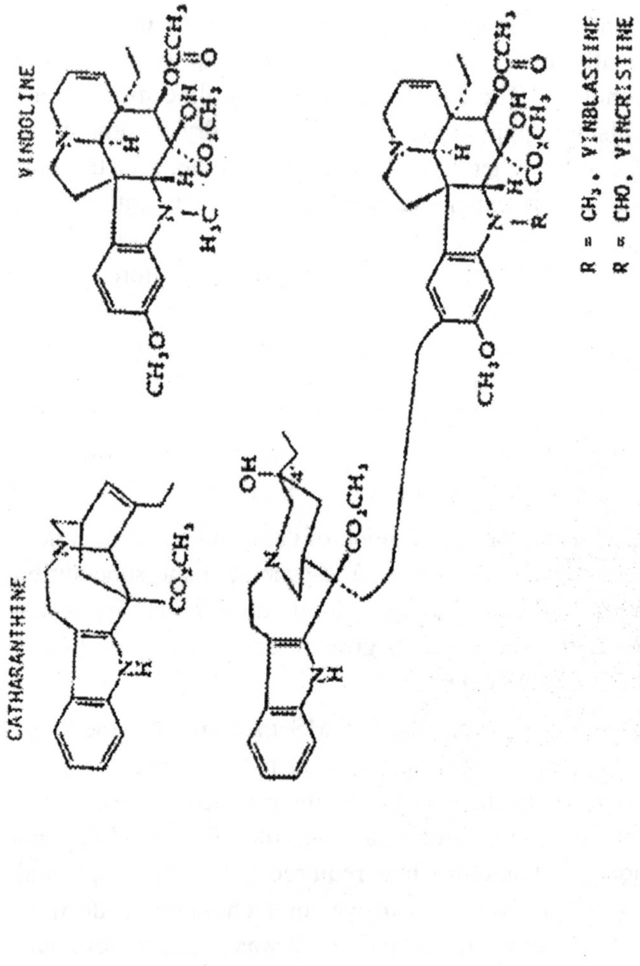

CATHARANTHINE

VINDOLINE

R = CH_3, VINBLASTINE

R = CHO, VINCRISTINE

Figure 2: Chemical Structures of Catharanthine, Vindoline, Vinblastine and Vincristine

sulphate, abscisic acid and NaCl on the production of catharanthine were significant .

Based on the conditions optimized by using flasks, Smart et al scaled up the cultures to 10, 30 and 100 L-air lift fermentors. When abscisic acid was added to the culture as an elicitor on the 7th day of cultivation, the final titer of catharanthine was raised to 85 mg/L in a 30 L fermentor. The second stage in this project, the Allelix's group tried to couple enzymatically or chemically catharanthine produced by the cell culture process with commercially available vindoline. As an enzyme source for the coupling, a crude preparation obtained by 70% ammonium sulphate precipitation from the cultured cells of C. roseus was used.

The reaction mixture containing both monomeric alkaloids, Tris buffer (pH 7.0) and the enzyme preparation was incubated at 30° C and for 3 hours. It was determined that the enzyme reaction gave various dimeric alkaloids including vinamidine, 3-(R)-hydroxyvinamidine and 3'4'-anhydrovinblastine. Leurosine and catharine, oxidized derivatives of anhydrovinblastine, were also detected in the early stages of the incubation. They found that $MnCl_2$ and either FAD or FMN stimulated the coupling. Although neither vinblastine nor vincristine was detected in the mixture, it was recognized that a substantial amount of anhydrovinblastine was formed as a major coupling product when an excess amount of sodium borohydride was added to the mixture after incubation.

In order to investigate properties of the coupling enzyme(s) it was partially purified with gel filtration and isoelectric focusing and five isozymes were obtained by Endo et al.. One of them had MW 15,000 and the other four had the same MW (37,000). All of these isozymes were shown to have peroxidase activity. Using the partially purified enzymes, anhydrovinblastine was formed with a conversion yield of about 50%. Formation of vinblastine from vincristine as detected by Goodbody et al. using a crude enzyme preparation obtained from suspension cultured cells of C. roseus.

The highest yield of conversion obtained was 13% from 0.13 mg anhydrovinblastine in 1 ml of the reaction mixture after 3 hours incubation at 30° C, Ph 7.0. During the course of these studies on coupling mechanisms, they found that ferric ion catalyzed the coupling reaction significantly in the absence of the enzyme. It is of interest that the products of the chemical coupling were not only anhydrovinblastine but also vinblastine. The yields of both alkaloids were 52.8% and 12.3%, respectively after 3 hours incubation at 30° C, pH 7.0. These products including catharanthine were analyzed by high resolution mass spectrometry as further confirmation of their identification. Circular dichroism confirmed that a-coupling exists between the 2 monomeric units of both vinblastine and vincristine produced either enzymatically or chemically.

This is a novel and an efficient process to produce an antitumor drug, vinblastine, and is likely to be applied commercially. The technology was transferred from the Canadian company to a Japanese company, Mitsui Petrochemicals Industry for further development. Hara et al. of Mitsui Petrochemical could increase the yield of catharanthine up to 150 mg/L in the MS medium supplemented with 1 mg/L NAA and 0.1 mg/L kinetin using the best producing cell line isolated from Allelix's cell line. The stimulating activity of NaCl and KCl on alkaloid production was also confirmed. Furthermore, the scientists of Mitsui employed high-cell density cultures and reported yields of catharanthine of 230 mg/L/week.

The yield of vinblastine by the chemical coupling reaction was also improved by the same group; addition of ferric chloride, oxalate, maleate and sodium borohydrate stimulated the yield of vinblastine from anhydrovinblastine up to 50%. Bede et al. also investigated the production of anhydrovinblastine. They employed a two-enzyme system containing horseradish peroxidase and glucose oxidase to catalyze the formation of anhydrovinblastine from catharanthine and vindoline. Although peroxidase requires hydrogen peroxide for the coupling reaction, its presence in excess in the reaction mixture may inhibit the reaction. But addition of glucose oxidase was used to allow the controlled, continuous production of hydrogen peroxide at low levels, minimizing oxidative reactions. Both enzymes were immobilized on Euperight C beads, an oxirane matrix, and the system was shown in catalyze the coupling reaction.

Taxol

Under the intensive NCl screening program of antitumor compounds in the U.S., Wall et al. began to isolate an active principle against KB cells from a tree, Taxus brevifolia in 1965. In 1969 pure taxol was first isolated and its chemical structure was disclosed in 1971. It is a diterpene amide and has shown against B16 mouse melanoma tumor, the MX-1 human mammary xenograft and CX-1 colon xenografts. The mode of action of taxol is rather unique because it stabilizes microtubles and inhibits depolymerization. The clinical trials begun in 1983 have shown positive results in the treatment of advanced ovarian cancer and breast cancer as well.

The FDA in the U.S. has approved taxol (generically known as paclitaxel) at the end of 1992; Bristol-Myers Squibb produces the drug for use in the treatment of ovarian cancer in patients who have failed to respond to other chemotherapies. Taxol is now being manufactured by extraction from the bark of wild-grown T. brevifolia trees. The demand for taxol will be undoubtedly increasing since it will be applied for other cancers including breast and lung cancers in the near future, but its supply is limited. Other related plants such as T. canadensis and T. cuspidata also contain taxol and other related compounds.

Generally the concentration of taxol in Taxus plants are very low. Therefore, harvesting Taxus trees for production of taxol commercially causes a serious problem in the U.S. from the environmental point of view. As alternative ways, plant tissue and cell cultures as well as chemical transformation processes from baccatin extracted from needles of the plant and the total chemical syntheses have been investigated in many research groups. In the U.S., Phyton Catalytic Inc. and ESCAgenetics announced a couple of years ago that they were establishing plant cell culture processes to manufacture taxol. The latter showed a photograph of a vial containing taxol powder produced by plant cell cultures.

Shuler et al. at Cornell University showed that a cell line of T. brevifolia in suspension provided by USDA-Agriculture Research Station and Phyton Catalytic, produced 3.9 mg/L taxol in the medium after 26 days of culture. The level of taxol was determined by reverse-phase HPLC. They found that fresh cell weight increased by 4-fold after a lag phase of about 4 days and that taxol was first seen at 13 days, and increased sharply after 20 days. It is of interest that all taxol produced was secreted into the medium, which was very unusual for plant cell cultures. Wickremesinhe and Artheca of the Pennsylvania State University established callus cultures and suspension cultures from T. brevifolia cv Repandens, T. cuspidata, T. media cvs. Hicksii and Densiformis.

Though some cell lines grew fast and their doubling times were 9-14 days, the levels of taxol were too low for commercial production. Induction of the hairy roots of Taxus plants has been tried. DiCosmo and his colleagues of the University of Toronto in collaboration of Nippon Oil Co. Ltd. in Japan described that they detected and identified taxol in the callus of T. cuspidata and its level was 0.02+0.005% in dry weight after 2 months in culture. Suspension cultures of T. cuspidata were also established from the callus cultures and subsequently immobilized onto glass fiber mats. The cells were maintained as immobilized cultures for 6 months. The level of taxol in the immobilized cells was 0.012+0.007% of the extracted dry weight.

Taxol is one of the most suitable and desirable plant products to plant cell culture research because the shortage of its supply and its high-value. Bristol-Myers Squibb is therefore, financially supporting many research groups to establish the process of cell culture production of taxol.

The root of Panax ginseng C.A. Meyer, a perennial herb, so-called "ginseng" has been widely used as a tonic and precious medicine since ancient times particularly in oriental countries including Korea and China. It is effective for gastroenteric disorders, diabetes and weak circulation, and has been used as an adjuvant to prevent various disorders, rather than a medicine to cure disorders. Thus, ginseng has been recognized as a miraculous medicine in preserving health and longevity. It has been known that the root contains various saponins and sapogenins. Among them, ginsenoside-Rb has a sedative activity, while Rg has a stimulative activities.

Although there are several species of ginseng, the commercially important species, P. ginseng grows in an area of 30-48° north latitude such as Korea. The cultivation of ginseng in the field requires four to seven years, and it is impossible to plant consecutively for 20 to 50 years but the demand for the plant has increased dramatically in the world, and its price has soared. These are reasons why many researchers have tried to produce ginseng cells through plant tissue cell cultures.

Furuya et al. at Kitasato University have studied P. ginseng callus tissues since the early 1970's. Meiji Seika Kaisha in Japan investigated the large-scale production of the cells established by Furuya using various types of fermentors. According to their patent published in 1973, crown gall calli, callus tissues and redifferentiated roots of P. ginseng were able to accumulate saponins and sapogenins known from the intact plant. The callus tissues and roots were cultivated on both MS solid and liquid media containing vitamins, sucrose, 2,4-D and suitable natural nutrients such as soybean powder or beef extract for several weeks at 25-28° C.

The concentrations of crude saponins in the callus (21.1%), in the crown gall (19.3%) and in the redifferentiated root (27.4%) were much higher than those in the natural root (4.1%). The saponins were found to contain ginsenoside-Rb and -Rg. To obtain high saponin-producing cells, mutagenesis was conducted using nitrosoguanidine and ?-ray as mutagens and a variant cell line of the drown gall induced by ?-ray irradiation was shown to accumulate 25.5% of saponins.

Glucose in the medium promoted cell growth in the initial stages of the fermentation and sucrose fed during the growth cycle stimulated the productivity of saponins. Although a higher NO2/NH3 ratio was favorable to the growth, it decreased the concentration of saponins in the cells. The growth was suppressed by moderate agitation, but the yield of saponins increased. The highest cell mass, 19 g/L on a dry weight basis was obtained using a 2 KL fermentor and the production rate of the cell mass was approximately 700 mg/L/day.

Rosmarinic Acid

Rosmarinic acid, or a-0-caffeoyl-3,4-dihydroxyphenyllactic acid has been found in the families Laminaceae and Boraginaceae. Rosmarinic acid and some related compounds were reported to have physiological or pharmaceutical activities. Oxidized rosmarinic acid was reported to show antithyrotropic activity and rosmarinic acid itself has been shown to effectively suppress the complement-dependent components of endotoxin shock in rabbits, however these compounds have not yet been used as commercial drugs.

Cultured cells of several plant species such as Coleus blumei, Anchusa officinalis and Lithospermum erythrorhizon were found to accumulate rosmarinic acid. It is of interest that production of the acid is stable, and high levels are produced. Ellis described that the production of rosmarinic acid appeared to be constitutively expressed in C. blumei cells, some of which had been in continuous culture for 10 years with no reduction in metabolite yield.

Dedifferentiated cell cultures of C. blumei and A. officinalis accumulated almost exclusively rosmarinic acid with the levels higher than in the intact plants when both types of cells were cultivated in a Gamborg and Eveleigh's B5 medium. The yield of the acid, before optimization, was about 1.4 g/L for C. blumei and 0.7 g/L for A. officinalis. Both species grew very well and reached up to 16 g-d.w./L within 30 to 40 hours of the culture period. Zenk et al. reported in 1977 that increasing the sucrose concentration in B5 medium up to 7.5% greatly stimulated both cell growth and rosmarinic acid formation in C. blumei cultures. The yield of the acid was approximately 3.6 g/L and the cell mass was 27 g-d.w./L. The yield shown seems to be one of the highest productivities of secondary metabolites in plant cell cultures.

Culture conditions and correlation between cell growth and production of rosmarinic acid were investigated extensively by De-Eknamkul and Ellis using both cell lines, for example NAA was the most favorable auxin to produce the product for A. officinalis cells, while 2,4-D was the most effective in C. blumei cultures. Alfermann and his colleagues in Germany have been investigating the biosynthetic pathway of rosmarinic acid and found two new enzymes, i.e., hydroxyphenylpyruvic acid reductase which catalyses the reduction of 4-hydroxyphenyl pyruvic acid to the corresponding lactic acids, and rosmarinic acid synthetase (caffeoyl-CoenzymeA:3,4-dihydroxyphenyllactic acid caffeoyl transferase), the enzyme transferring the caffeoyl moiety from caffeoyl-CoA to 3,4-dihydroxyphenyllactic acid. This is the crucial enzyme in rosmarinic acid biosynthesis forming the ester linkage between the caffeic acid moiety and the 3,4-dihydroxyphenyllactic acid moiety.

To confirm localization of rosmarinic acid accumulated and its biosynthetic enzymes in the cells, the same group prepared protoplasts from C. blumei cultured cells. Rosmarinic acid as well as the enzymes were mainly found in vacuoles. They also purified some of the enzymes. Recently Mizukami and Ellis isolated three isoforms of tyrosine aminotransferase (TAT) from Anchusa officinalis cell suspension cultures and indicated two (TAT-1, TAT-4) out of three isoforms were involved in the synthesis of rosmarinic acid. They suggested that TAT-1 controls conversion of tyrosine to 4-hydroxyphenyl pyruvate and TAT-4 acts by participating in the formation of tyrosine and phenylalanine via prephenate.

Determination of the biosynthetic pathway of secondary products will undoubtedly contribute to the further improvement of producing cells using recombinant DNA technology. This technology should be applied more extensively in the future. Ulbrich et al. of A. Nattermann & Cie. GmbH in Germany investigated large-scale production of rosmarinic acid by a C. blumei cell culture process. In order to increase oxygen supply in the medium and to operate at high cell density, they designed a special module spiral stirrer. It was built of six modules, and a special end module each consisting of one plain metal ring (spiral blade) fixed to the stirrer shaft with two spokes and two rings.

The rotation speed of the module spiral stirrer ranged between 50-100 rpm without significant cell damage. Standard rotation speed was set at 100 rpm. The inoculum of C. blumei cells was cultivated in a fed-batch process and then about 30-50% culture broth was transferred from the seed fermentor to the production fermentor with sucrose solution (50 g/L) as the production medium. Using this procedure the yield of rosmarinic acid increased to 5.5 g/L or 910 mg/L/day, representing 21% of the dry weight. According to Ulbrich's experiments, the process needs only 14 days to produce in total 200 g of 97% purity of rosmarinic acid in two parallel production batches from one inoculum fermentor (32 L working volume). It is not necessary to separate the cell mass from the growth medium, and they would recycle it as a part (30% v/v) of the simple and cheap production medium (sucrose 50 g/L). It is hoped that some commercial utility for rosmarinic acid will be found in the future.

Arbutin

Arbutin is widely distributed in various plants of the Ericaceae such as Arctostaphylos uva-ursi and Vaccinium vitisidaea. The level of arbutin in the bark of Pyrus communis reaches up to 28%. Arctostaphylos uva-ursi has been used as a urethal disinfectant and its major active principle, arbutin, was shown to suppress the synthesis of melanin in human skin. A Japanese cosmetic company, Shiseido, has developed arbutin as an additive for the company's product lines because of its preventive activity toward pigmentation of skin.

Although arbutin is commercially available by a chemical process, researchers of Shiseido have investigated an alternative process using plant cell cultures including biotransformation. Tabata et al. showed that cultured cells of Datura innoxia had a remarkably high capability for glucosylation of hydroquinone to form arbutin, and hydroquinone was totally converted to arbutin within 10 hours after administration. The glucosylation is catalyzed by an enzyme, uridine diphosphate glucose (UDPG)-hydroquinone glucosyltransferase.

Yokoyama et al. selected C. roseus cells as a producer of the enzyme since arbutin was formed efficiently when hydroquinone was added into the suspension culture. They have optimized various components in Linsmaier-Skoog's basal medium for production of arbutin and found that higher levels of sugars such as sucrose in the medium, up to 6%, gave higher yields of arbutin. Concerning this sugar effect, Yokoyama suggested that sucrose was a scavenger, and therefore it protected cultured cells from the damaging activity of hydroxy radicals of the substrate. This was also supported by their finding that some antioxidants such as ascorbic acid, gallic acid, cysteine, tannin and phytic acid increased the level of arbutin.

One of the high producing cell lines, C. roseus strain B was cultivated in a 5 L jar fermentor equipped with modified paddle-type impellers and spargers of porous sintered metal which prevents to some extent the adhesion of cells to the inner surface of the fermentor. Glucose was used as a carbon source instead of sucrose for economic reasons for mass production of arbutin. In order to increase the cell density in the medium, 10 times concentration of the medium components was fed to the medium during the fermentation.

It was prerequisite to keep the level of hydroquinone in the medium as low as possible to avoid damage of the cells by the substrate. Therefore, hydroquinone was fed at 6 mM in the beginning of the fermentation and after its concentration had decreased to around 0 mM in the medium, they started to feed hydroquinone continuously at 1.4 mmol per hour. Under these conditions, 9.2 g/L of arbutin which corresponded to 45% of dry cell weight was obtained in 3 to 4 days after administration of hydroquinone. Similar yields was also obtained using larger scale fermentors.

Since the production continued until cell death, arbutin was accumulated extracellularly and its concentration reached approximately 1% in the culture filtrate at the end of the production phase. Therefore, arbutin was easily extracted from the filtrate. The total cultivation period was approximately 18 days including 2 weeks for high density cell culture and 3 or 4 days for the biotransformation process. Biotransformation is one of the most feasible processes in terms of industrial application of plant tissue and cell cultures. It is advantageous that the yield of arbutin is high and the cost of hydroquinone is inexpensive enough as a substrate, accordingly the author believes that the process will be employed for commercial manufacturing arbutin in the near future.

AGRICULTURAL DRUGS

Plant Viral Inhibitors

A large number of chemically synthesized compounds and natural molecules have been examined for their inhibitory effects on plant viruses and some of them have potent activity as protectors against virus infection. In order to screen high producing cultured cells of plant viral inhibitors, a variety of callus extracts were examined the inhibitory activity towards tobacco mosaic virus (TMV) infection using a tobacco disc method by Misawa et al. of Kyowa Hakko in Japan. Among these extracts Phytolacca americana callus was selected as the most potent producer of the inhibitor.

The level of the inhibitor accumulated in the suspension cultured cells and reached maximum level on the 9th day of culture using MS medium containing 1 mg/l 2,4-D. The cell suspension of P. americana was homogenized and the supernatant was diluted up to 100 times with water. The diluted solution was found to inhibit TMV infection on tobacco and tomato plants significantly. Further studies in the same laboratory used Agrostemma githago, a more potent producer of the plant virus inhibitor. The growth of suspension cultured cells of A. githago was somewhat faster than P. americana.

Active principles of P. americana and A. githago were isolated with Column-lite chromatography and electrofocusing. At least four basic proteins were obtained from P. americana cells whose molecular weights were 1.10X104 to 3.15X104 Among them the highest molecular weight component contained sugars in the molecule. On the other hand, only one basic protein was isolated as the principle from A. githago culture and its molecular weight was 2.5X104. The proteins obtained from P. americana have been widely investigated because of their activities to HIV.

Ikeda et al. of Japan Tobacco Inc. also screened various plants and selected Mirabilis jalapa as a producer of an anti-plant virus protein. The callus was induced from leaves of M. jalapa and its suspension cultured cells established were found to accumulate the protein intracellularly. Optimization of the production and the cell growth as well as selection of high producing cell lines were conducted extensively. One of the lines produced 95 mg/L of the protein in the optimized medium based on MS medium on the 7th day of the cultivation. Its molecular weight was 24 KD and the amino acid sequence was determined which had 24% homology with a ribosome-inactivating protein, ricin D-A chain.

FOOD ADDITIVES

Pigments

Shikonin compounds

Shikonin and its derivatives such as acetyl shikonin and isobutyl shikonin accumulated in roots of Lithospermum erythrorhizon Sieb. & Zucc. are reddish purple pigments and have been used in traditional dyeing. The plant has also been used as a herbal medicine. Because of a shortage of this plant, Fujita et al. at Mitsui Petrochemical in Japan investigated mass cultivation of L. erythrorhizon cells to produce shikonin compounds. Using the cell line established by Tabata's laboratory of Kyoto University, they optimized its culture conditions extensively to increase the level of the products using flasks and various types of fermentors including a rotating cylindrical fermentor designed by Tanaka et al. of Tsukuba University.

Fujita et al. found that L. erythrorhizon produced shikonins in White's medium but the cell growth was poor in the same medium. On the other hand, Linsmair-Skoog's (LS) medium was recognized to support the growth but not shikonin production. Therefore, they used a two-stage culture for mass production of shikonin compounds. Namely, to proliferate the cells, LS medium was used at the first stage of the fermentation, and then the cells were transferred into White's medium for production of shikonin compounds. In order to improve the yields further, optimization of components in both media was carried out extensively, and MG-5 and MG-9 media were established,

Since most cultured cells in liquid and solid media occur as aggregates, selection of high-producing cell lines from the aggregated cells is not effective and labour-intensive. The Mitsui group prepared protoplasts from the cultured cells with appropriate enzymes and selected high shikonin compounds-producing protoplasts using a cell sorter. The selected protoplasts were generated to cell lines and cultivated in suspension. From 48 cell lines, they obtained a cell line having 1.8 fold the productivity of the parent line. The cell line showed stable production of shikonin compounds.

To produce the compounds more efficiently, the same group attempted to employ a high-cell density culture process in the second stage of the two-stage cultures. By feeding the nutrients into M-9 medium, the level of cell mass increased and that of the compounds produced increase twice as much as without feeding. Shikonin and its derivatives are being manufactured commercially by the company. A major application of the pigment is for lipsticks. Shimomura et al. established a hairy root culture of L. erythrorhizon with Agrobacterium rhizogenes.

The hairy root culture did not produce shikonin on solid MS medium but produced the pigment in the root culture medium and also secreted it into the medium. Addition of absorbents; XAD-2, XAD-4, charcoal and so on increased the concentration of shikonin produced. The roots were cultivated in a 2 L air-lift type fermentor connected to a XAD-2 column (25 g) through a peristatic pump and 5 mg/day of shikonin was continuously produced during a period of more than 220 days.

Anthocyanins

Anthocyanins are the large group of water-soluble pigments responsible for many of the bright colours seen in flowers and fruit. They are composed of an aglycone (anthocyanidin) and more than one sugar moiety and normally change colour over the pH range due to the existence of four pH-dependent forms. Thus, at low pH they are red and at higher pH value (over 6) they turn blue. They are commonly used in acidic solutions in order to impart a red colour to soft drinks, sugar confectionary, jams and bakery toppings.

The major source of anthocyanins for commercial purposes are grape pomaces and waste from juice and wine industries, but other potential sources have been investigated. Crude preparations of anthocyanins are used extensively in the food industry and it has been claimed that pure anthocyanins are priced \$1,250-2,000/kg, but crude materials are rather inexpensive. Commercial exploration of cell cultures for anthocyanins therefore, has not been tackled seriously. Although there have been many papers describing the production of anthocyanins using cultured cells of various plant species, most of them seem to use an anthocyanin-producing cell line as a model system for secondary product production because of their colour which allows production to be easily visualized.

Among them, Yamamoto et al. of Nippon Paint Co. in Japan have studied production of anthocyanins intensively. They induced callus from Euphorbia millii leaves on MS medium containing 2,4-D, NAA, natural sources such as malt extract and yeast extract. As a major component in the callus, they identified cyanidin-3-arabinoside. The callus consisted of cell aggregates was cut to small pieces and cultivated on solid agar media. High producing cell aggregates were selected visually and they were transferred to fresh agar media. This procedure was repeated 28 times and one of the cells was determined to produce 1.32% d.w. anthocyanins in the cells.

The levels of the pigments in flowers and leaves were 0.28% and less than 0.01%, respectively. They also established suspension cultures of E. millii. Accumulation of anthocyanins was enhanced by a high osmotic potential in Vitis vinifera L. (grape) cell suspension cultures (249). They added sucrose or mannitol in the medium to increase the osmotic pressure and found the level of anthocyanins accumulated was increased to 1.5 times, 550 μg/10 cells.

Safflower yellow

This yellow pigment obtained from the floret of the safflower plant (Carthamus tinctorius L.), is also known as Mexican saffron or American saffron, although it has no relation to genuine saffron. The major pigment is carthamin, which exists at levels of up to 30% in the flowers, and there is also a red pigment in concentrations of about 0.5%. Carthamin is the quinoid form of isocarthamin, the glucoside of 2',3',4',6'-tetrahydro-chalcone.

Safflower yellow is not approved for use in the U.S. or in the E.E.C., but regulations do permit its use in Japan. It is stable to heat and light and is used in baked goods and beverages. The production of carthamin from safflower callus cultures has been described by Kibun Co. in Japan. The callus was obtained from flower bud explants and could also be put into suspension. Medium optimization has been performed. The production of alpha-tocopherol (the tocopherol with the highest vitamin E activity) has been described for safflower cultures. Selection with various media components and precursor feeding experiments have enhanced the production.

Kusaka et al. found that addition of cellulose, chitin or chitosan increased production of the red pigment. These polysaccharides appeared to show an eliciting activity. Addition of 1 mM D-phenylalanine and removal of Mg alone or both Mg and Ca from the culture medium also increased the production. Plant cell cultures cannot presently be used for the production of these metabolites, given the high cost of the technology and the low value products, ie. $50-$80 and $48 for tocopherol and carthamin respectively.

Saffron

Saffron is the name of the spice which is made from the stamens of Crocus sativus and are prized for their use as a flavoring and colourant. The stigma of the plant contains crocin (yellow pigment), safranal (a fragrance) and picocrocin (bitter substance). The plant is grown mainly in Spain and India and it requires about 30,000-35,000 hand-picked blooms to produce 1 lb of dry saffron. Crocin, being a glycoside, is water-soluble and is not soluble in oils and fats.

Saffron is sensitive to pH changes and is unstable towards light and oxidative conditions, but it is moderately resistant to heat. It is used in baked goods, soups, meat and curry products, cheese, confectionary and as a condiment for the rice of Spanish and Indian foods. Saffron is also reputed to have medicinal value for stomach ailments.

The very high value of the product is due mainly to the fact that the life of the flowers is very short, making harvesting difficult. Thus, this is an ideal target for plant tissue cultures. Ajinomoto of Japan have approached this problem through the propagation of saffron stigma-like structures in vitro. Further studies showed that crocin and picrocrocin were present and, after heat treatment (as done with field-grown stigmas), safranal was produced. The composition of these phytochemicals corresponded with that of similarly-treated young, intact stigmas.

Madder colorants

Rubia tinctorum (Rubiaceae) is a perennial plant, madder, originated from the coastal regions of the Mediterranean and its roots have been used as red dyes in western Europe. The major components in the pigment are alizarin, purpurine and its glycoside, ruberythric acid. Pure alizarin is an orange crystal and is soluble at 1 part to 300 in boiling water and in other solvents. The R. tinctorum pigment, so-called madder colorant, shows a yellow color in acidic to neutral pH and tends to be reddish with increasing pH. It is highly resistant to heat and light which is favorable to food industry.

The callus of R. tinctorum was induced from the root of a germ-free plant by Odake et al. of San-Ei Chemical Industries in Japan and was grown on LS agar medium containing 2,4-D and 0.2% gellan gum. Through the selection of high-producing cell lines and successive transfers, yellow pigment-producing cells were obtained which were then transferred into a liquid LS medium containing 10-6 M 2,4-D, 10-6 M kinetin and 3% sucrose. After 21 days cultivation in a 100 L jar fermentor, approximately 1.5 g of the pigment was extracted.

In order to remove auxins such as 2,4-D and IAA which are not desirable for food industry, a hairy root culture of R. tinctorum was established by the same group using Agrobacterium rhizogenes. They used a 5 mm diameter disc obtained from a leaf of R. tinctorum. It was incubated with cells of A. rhizogenes in suspension. After 78 hours at 25° C in the dark, the leaf disc was transplanted to a hormone-free solid LS medium containing claforan (0.05%), sucrose (3%) and gellan gum (0.2%). Fourteen days later, hairy roots were found to produce the pigment. Using a 100 L bubble-column type jar fermentor, the hairy roots were cultivated for 21 days and approximately 800 mg of the pigment was obtained.

Miscellaneous

Chicle

Chicle is the most important raw material in chewing-gums and is made from the latex of Achras sapota Linn. Chicle contains approximately 60% of resin and 15% of rubber. The resin consists of lupeol, a-amyrin and ß-amyrin, and the rubber fraction contains cis- and trans-1,4-polyisoprene, however the biosynthetic pathway of chicle and its regulation mechanisms have not been elucidated although components of chicle are suggested to be synthesized through the mevalonic acid pathway.

The callus induced from young shoots on a Linmaier-Skoog's medium was shown to produce lupeol acetate, palmitate and stearate neither a,ß-amyrin nor rubber. On the other hand, the cells cultivated in suspension produced triterpenes as well as phytosterols such as campesterol, cholesterol, ß-sitosterol and stigmasterol. In spite of many efforts for optimization of the culture conditions in suspension, the yield of triterpenes was not high enough for commercial application of the cell culture, therefore the researchers induced callus tissue of Dyera costulata and Couma macrocarpa Bars Rodr., both of which were known to produce latexes for chicle as well. The callus tissues of both plants produced triterpenes at higher levels than those in the intact plants, respectively, but not the rubber.

Mucilage

Polysaccharides produced by Astragalas gummifer have been used as additives of ice cream and edible dressings. Isa et al. of Q.P. Corp. in Japan used a hairy-root culture technology in order to establish an alternative method of production of mucilages of A. gummifer because of high cost and unstable supply of natural gums.

A. rhizogenes was inoculated in the stem of the in vitro plantlet of A. gummifer, which was incubated for approximately 14 days. Hairy root tips induced at the inoculated site were excised and cultivated on a hormone-free MS medium solidified with 0.2% Gelrite containing 500 mg/L Clafovan, a cefotaxim antibiotic. After successive transfers on the solid medium, the hairy roots were transferred into 30 ml of hormone-free liquid medium in a 100 ml Erlenmyer flask and cultivated at 25° C in the dark with constant agitation at 60 r.p.m. The cells transformed by A. rhizogenes were found to produce opines as well as several types of water soluble mucilages. The composition of monosaccharides such as glucose, arabinose, galactose and xylose in the mucilages obtained from different hairy root lines varied widely, but the chemical structure of mucilages in the mother plant is much more complicated.

Hernandulcin

Hernandulcin is a sesquiterpene compound having strong sweet taste isolated from Lippia dulcia (Verbenaceae), but it has not yet been approved by FDA in the U.S.A. Sauerwein and Shimomura induced hairy roots of this plant using A. rhizogenes. The roots cultivated in MS liquid medium supplemented with 2% sucrose under 16 hr/day light accumulated 0.25 mg/g-d.w. of hernandulcin. Addition of 0.2-10 mg/L chitosan into the medium increased its level up to 5 times. The axenic shoot culture of L. dulcia on MS solid medium containing 2% sucrose was shown to produce a high concentration of hernandulcin, 2.9%-d.w..

There are still a number of commercially important products which are being extracted from mass produced field-plants. In order to circumvent various problems caused from these processes as the author indicated in the Introduction, plant cell culture technology has been expected to be an efficient and useful tool. For more than 30 years, many researchers have investigated plant cell cultures for production of a variety of phytochemicals; however, in spite of their many efforts only two products such as shikonins and ginseng cells are so far being manufactured commercially.

The reasons why this technology has scarcely been applied in industry are; low yield of plant metabolites, unstable producing ability of cultured cells and their slow growth rate. Therefore, at present the plant cell culture is not a cost-effective technology. In particular, any process using plant cell cultures is not favorable if the desirable products can be easily manufactured by chemical- or microbial fermentation methods. However, a variety of scientific strategies have been investigated for improving the production ability of cultured cells and substantial progress has been made.

Undoubtedly, some plant metabolites are likely to be manufactured through plant cell cultures. To overcome barriers hindering industrial application of plant cell cultures, however, it is required not only to follow the approaches reviewed, but also to conduct more fundamental research, including elucidation of biosynthetic pathways of many useful secondary metabolites in plants and of the regulation mechanisms in their biosyntheses.

BIBLIOGRAPHY

Agblevor, F.A.; Rejai, B.; Evans, R.J.; Johnson, K.D., "Pyrolytic Analysis and Catalytic Upgrading of Lignocellulosic Materials by Molecular-Beam Mass Spectrometry." *Presented at Energy from Biomass and Wastes XVI*, Orlando, FL, 2-6 March, 1992.

_____________, "Compositional Analysis of Biomass Feedstocks and Chemical Analysis and Testing Standard Procedures." American Society for Materials and Testing (ASTM) E-48 Committee on Biotechnology meeting in Washington D.C., October 27-28, 1993, Gaithersburg, MD, 1993.

_____________, Czernik, S.; Davis, M.; Wang, D., "Impact of Storage Conditions on the Production of Hydrocarbon Fuels from Biomass Feedstocks." *Presented at the American Institute of Chemical Engineers (AIChE) Annual Meeting*, St. Louis, MO, November 7-12, 1993.

Bok, J.; Goers, S.; Eveleigh, D. "The Cellulase and Xylanase Systems of *Thermotoga neapolitana*." *Enzymatic Conversion of Biomass for Fuels Production*. Edited by M.E. Himmel, J.D. Baker, and R.P., American Chemical Society Symposium Series 566. Washington, DC: American Chemical Society, pp. 54-65, 1994.

Bradshaw, H.D., Jr.; Villar, M; Watson, B.D.; Otto, K.G.; Stewart, F.; Stettler, R.F., "Molecular Genetics of Growth and Development in *Populus*. III. A Genetic Linkage Map of a Hybrid Poplar Composed of RFLP, STS, and RAPD Markers." *Theoretical and Applied Genetics* 89(5):551-558, 1994.

Bulls, M.; Holmes, J., *Dilute Acid Pretreatment and Simultaneous Saccharification Fermentation (SSF) of Hybrid Poplar and Switchgrass QA/QC Plan*. Golden, CO: *National Renewable Energy Laboratory*, 1995.

Couto, L.; Betters, D.R., "An Overview of Eucalypt Plantations in Brazil." *California Eucalyptus Grower* 9:6-9, 1994.

Dale, M.C.; Tyson, G.; Zhao, C.; Lei, S., "The Xylan Delignification Process for Biomass Conversion to Ethanol." *Appl. Biochem. Biotechnol,* 1995.

Elander, R.T.; Hsu, T., "Processing and Economic Impacts of Biomass Delignification for Ethanol Production." *Appl. Biochem. Biotechnol.* Vol. 51/52, pp. 463-478, 1995.

Graham, R.L.; Downing, M., *Potential Supply and Cost of Biomass from Energy Crops in the TVA Region*. ORNL-6858, Oak Ridge, TN: Oak Ridge National Laboratory, 1995.

Hoffman, W.; Beyea, J.; Cook, J.H., "Ecology of Agricultural Monocultures: Some Consequences for Biodiversity in Biomass Energy Farms." *In Proceedings of the Second Biomass Conference of the Americas: Energy, Environment, Agriculture, and Industry*, Portland, OR, August 21-24, 1995, pp. 1618-1627, 1995.

Jeffries, T.W.; Davis, B., *Annual Progress Report: Genetic Engineering to Develop Improved Xylose-Fermenting Yeasts*. Golden, CO: National Renewable Energy Laboratory, 1994.

Kadam, K.L., *Power Plant Flue Gas as a Source of CO_2 for Microalgae Cultivation: Technology and Economics of CO_2 Recovery and Delivery*, 1995.

Wiselogel, A., (ed.), *Terrestrial Biomass Interface Feedstock Project Guidance Committee Charter,* 1995.

______________, "Production of Biomass Crops on CRP Land." *National Association of Conservation Districts*, New Orleans, LA, February 8, 1995.

INDEX